KW-054-260

MECHANICAL ENGINEERING THERMODYNAMICS:
A Laboratory Course

M. A. PLINT
B.Sc. (Eng.), Ph.D., C.Eng., F.I.Mech.E.

and

L. BÖSWIRTH
Prof., Dr Techn.

MORAY COLLEGE OF FURTHER
EDUCATION   ▪   LIBRARY

CHARLES GRIFFIN & COMPANY LTD
London

Quis separabit nos

CHARLES GRIFFIN & COMPANY LIMITED
Registered office:
16 Pembridge Road, London W11 3HL
England

Copyright © English edition, 1986
Charles Griffin & Company Limited, London
Copyright © German edition, 1977
Hermann Schroedel Verlag KG, Hanover

All rights reserved. No part of this publication may be
reproduced or transmitted in any form or by any means,
electronic or mechanical, including photocopying, recording,
or by any information storage and retrieval system,
without permission in writing from
Charles Griffin & Company Limited.

This edition of the work entitled *Technische Wärmelehre:
Ein Laboratoriumslehrgang*, by Leopold Böswirth and
Michael Alexander Plint, is published under licence from
the originating publisher
Hermann Schroedel Verlag KG, Hanover.

British Library Cataloguing in Publication Data

Plint, M. A.
   Mechanical engineering thermodynamics: a
   laboratory guide.
   1. Thermodynamics
   I. Title   II. Böswirth, L.   III. Technische
   Wärmelhre. *English*
   536'.7'024621   TJ265

   ISBN 0–85264–276–8

Typeset in Great Britain by
Latimer Trend & Company Limited, Plymouth
Printed & bound in Great Britain by
Redwood Burn Limited, Trowbridge, Wiltshire

£19.95

University of the
Highlands and Islands
Moray College

Learning Resource Centre
The Library, Moray Street, ELGIN. IV30 1JJ

Tel: (01343) 576206
e-mail: mc.library@moray.uhi.ac.uk

This book is due for return on or before the last date stamped below

MORAY COLLEGE OF FURTHER EDUCATION · LIBRARY

MECHANICAL ENGINEERING THERMODYNAMICS:

A Laboratory Course

By the same authors
Fluid Mechanics: a Laboratory Course

Contents

Preface

There is general agreement that laboratory work of some kind should form part of the training of all engineers, whether at professional or technician level. The design of suitable experimental work in the field of thermodynamics presents peculiar difficulties, partly because of the nature of the subject itself and partly as a consequence of its historical development. The traditional "engine test", while certainly not devoid of value as helping to give the student a "feel" for engineering and a necessary confidence in the handling of noisy and fast-running machinery, is of limited value as a source of illumination of the theory of the subject. This is partly because of a degree of mismatch between the contents of classical thermodynamics and the body of knowledge associated with its application. The petrol engine, for example, has been the subject of more development work than any other machine; problems associated with mixture formation, flame propagation and combustion, the shape of the combustion chamber and the design of the inlet and outlet passages have occupied engineers for three-quarters of a century and are nowhere near a final solution. The contribution made by classical thermodynamics to the solution of these problems has, however, not been great.

A further problem arises from the fact that classical thermodynamics could more accurately be described as thermostatics: in the world of the classical thermodynamicist nothing happens, or it happens infinitely slowly. This makes it difficult for the student to appreciate the relevance of the theory to processes taking place in the real world. The authors have attempted to rectify this situation in some small measure by reversing the usual order of experimental work and dealing with heat transfer before coming on to work involving heat engines proper. This not only emphasizes the importance of processes taking place in time but defers consideration of problems involving the more difficult concepts associated with the Second Law of Thermodynamics until a later stage in the laboratory course. In their choice of experiments the authors have tried to strike a balance between those dealing with purely theoretical aspects of the subject and those relevant to practical applications. The book follows an earlier volume [11] dealing with the subject of fluid mechanics and assumes a similar pattern. The primary purpose is to provide the essential material for a course of laboratory work in thermodynamics in a form designed to reduce the teacher's preparatory labour to manageable proportions.

It is intended that this book should be suitable for use in two different ways: either with full participation by the lecturer, or independently by the student with only limited instruction and supervision at a practical level by a laboratory technician or assistant. The structure of the book is arranged to facilitate this. In each of Chapters 2 to 9, after a summary of the relevant theory, a description is given of the experiments and of suitable apparatus with which they may be carried out. The authors make no apology for referring to specific equipment at this point, since they believe that the value of their work would be greatly reduced if they merely descended to generalities. The difficulties encountered in laboratory work are nearly always matters of practical detail that do not come to light until the experiment is actually carried out. In this connection it may be emphazied that every experiment described has been performed by one or other of the authors using the apparatus in the book. It is not, of course, intended that identical apparatus should necessarily be used by the reader; however, the fact that the experiments described have actually been performed is a guarantee that with adequate equipment no serious snags are to be expected.

A brief account of the experimental procedure follows. It is not intended to "spoon feed" the student to an excessive degree; nevertheless, the major pitfalls are pointed out and the student is warned of the errors and omissions that so often invalidate experimental results, both in the teaching laboratory and at a more advanced level.

Following the section on experimental procedure comes one describing a typical test and discussing salient features. Actual measurements are used as examples, but in general only one observation is fully worked out so that mere copying of results from the book is discouraged. Finally, suggestions are made for further experimental work forming a natural extension to the theme of the chapter. It is hoped that these suggestions will form a useful source of ideas for project work of the kind frequently undertaken by students towards the end of their course. Notes for guidance in connection with the general experimental methods to be adopted are given in the Appendix.

The range of experiments described is suitable for inclusion in a course at degree level; with perhaps certain omissions and simplifications, they are equally appropriate for courses at technician level such as those for the Higher Technical Certificate and Higher National Diploma in Engineering.

The authors' treatment of the theoretical aspects of the subject may be considered as in some respects unconventional. No attempt has been made to reproduce the traditional extensive mathematical treatment of classical thermodynamic theory, such as may be found in many standard texts [1,2]. They have, on the contrary, attempted to reduce the amount of mathematics but to emphasize at every point the connection between thermodynamic abstractions and phenomena taking place in the real world. Many references are made to the kinetic theory of heat since this can, in the authors' opinion, provide numerous valuable insights.

A word as to the origins of the book. It has arisen as a result of a year-long collaboration between the authors at the Höhere Technische Bundeslehr und Versuchsanstalt in Mödling, Austria. One author is a permanent member of the staff of this College, the largest engineering school of its kind in Austria, while the other author has for many years been active in the development of engineering laboratory apparatus in England. Two versions of the book, one English and one German, have been prepared, and it is hoped that a by-product of the book may be some small contribution to a mutual understanding of the special features and individual virtues of the Continental and British systems of engineering education.

The authors gratefully acknowledge the advice and help of a number of distingushed teachers in England, Germany and Austria; in particular of Professor George Jackson of Brunel University, Dr E. Glaister, lately Reader in Thermodynamics at Imperial College, London and Paul Minton, Senior Lecturer in Hydraulics at Imperial College, also that of Martin Griffin, who prepared the numerous illustrations.

Finally, they record their thanks to the late Professor Dr H. Schlöss of the Austrian Bundesministerium für Unterricht und Kunst, without whose help and encouragement this book would never have been written.

M. A. PLINT   L. BÖSWIRTH

Wokingham/Mödling
May 1986

Principal Symbols

In accordance with British Standard 1991: Parts 1 (1976) and 5 (1961); Letter Symbols, Signs and Abbreviations

a	thermal diffusivity
A	surface area; non-flow availability function
B	steady flow availability function
c	specific heat; mean velocity of molecule
c_p	specific heat at constant pressure
c_v	specific heat at constant volume
C_d	coefficient of discharge
C_D	drag coefficient
C_f	surface friction coefficient
d	diameter
D	drag force
e	excess air ratio
E	energy
F	force
$\dot{F}$	fuel flow rate
g	gravitational acceleration
(Gr)	Grashof Number
h	height; height of fluid column; specific enthalpy
h_0	barometric pressure
H	enthalpy
I	electric current
J	mechanical equivalent of heat
$k_1, k_2, \ldots$	constants
K	Boltzmann constant
(Kn)	Knudsen Number
l	characteristic length; latent heat
m	mass
$\dot{m}$	mass flow rate
M	moment
n	rev/min
(Nu)	Nusselt Number
p	pressure
p_0	barometric pressure
P	power

(Pr)	Prandtl Number
q	quantity of heat per unit mass or area
$\dot{q}$	heat flow rate per unit mass or area
Q	quantity of heat
$\dot{Q}$	heat flow rate
$\dot{Q}_{gr}$	gross or higher calorific value
$\dot{Q}_{net}$	net or lower calorific value
r	radius
R	gas constant
(Re)	Reynolds Number
s	specific entropy
S	entropy
t	time; temperature °C
T	temperature K
t_∞, T_∞	free stream temperature
u	specific internal energy
U	internal energy
v	velocity; specific volume
V	volume; electrical potential
w_s	shaft work
W	work
$\dot{W}$	rate of work or output
x	dryness fraction
z	height above datum
α	heat transfer coefficient; angle
β	coefficient of cubical expansion
γ	ratio c_p/c_v
δ	boundary layer thickness
ε	emissivity; compressibility coefficient
η	efficiency, dimensionless; distance from wall
η_C	Carnot cycle efficiency
λ	thermal conductivity; mean free path; air/fuel ratio
μ	dynamic viscosity
v	kinematic viscosity
ρ	density
σ	Stefan–Boltzmann constant
τ	time constant; shear stress
τ_0	shear stress at wall
φ	relative humidity
ω	angular velocity; specific humidity

Units

m	metre		W	watt
s	second		K	degree Kelvin
kg	kilogram		°C	degree Celsius
N	newton		V	volt
J	joule		A	ampere

1

Purpose of the Book and Method of Use

The purpose of this book is to give in readily accessible form sufficient information to enable the student to complete, with no wasted effort, a range of experimental work covering the fundamentals of engineering thermodynamics. The book also contains a brief statement of the essential theoretical material. This statement differs somewhat from that given in typical standard texts, in placing less emphasis on the more abstract aspects of the subject and more on the practical implications. It is hoped that the reader who is already familiar with the standard treatment may gain new insights as a consequence of this alternative approach.

The layout of the chapters is as follows. Each starts with a brief summary of the background theory forming the basis for the experiments that follow; this is not intended to be a comprehensive exposition but rather an *aide-mémoire* to call freshly to mind the essential principles, which will already have been studied in the lecture theatre. The teacher using this book will, it is hoped, find that this summary forms an adequate set of lecture notes for an introductory talk which, it is suggested, should take place a few days before the laboratory session. Alternatively, the student should study this section himself and should refer back to the relevant lecture notes and textbooks, again before coming into the laboratory. Some material which is of a more advanced nature and not essential to an understanding of the experiment is given in small type. It may be omitted by the more elementary student.

Next follows a description of the experiment(s) and of the apparatus that was used by the authors themselves in carrying them out. It is not, of course, necessary that identical equipment should be used by the reader. It is recommended that the student should inspect the apparatus some time before the class is due to take place; for example, towards the conclusion of the previous laboratory period.

Next follows a section describing the specific theoretical basis for the particular experiment, defining the various constants and dimensionless groups and outlining the mathematical processes by which the results are calculated. The section on experimental procedure should be studied with care, and supplemented by an examination of the apparatus and of any more detailed instructions that may be provided in the laboratory.

At this point the student should be in a position to carry out the experimental work. It will be found that in general a group of three or four students will be adequately occupied for a laboratory period of from two to three hours in taking a set of readings, making a preliminary plot of the results and, so far as possible, completing all necessary calculations.

Following the description of the experimental procedure, each chapter gives one or two typical results in order to illustrate the method of working them out. It has not been considered appropriate to give an extensive discussion of the results; this is or should be part of the student's task if he is to obtain maximum benefit from his laboratory experience. However, various suggestions as to points arising from the observations that invite discussion have been made.

Finally, each chapter closes with suggestions as to further possibilities for more advanced experimental work, in many cases using the same apparatus, and it is hoped that these may be of use when planning more advanced project work. It is recommended that the various experiments should be carried out as soon as possible after the corresponding topic has been encountered in the lecture theatre, thus reinforcing the impact of the lectures to the maximum degree.

An exception to this pattern is presented by Chapter 2, which deals with the fundamentals of thermodynamics in a simple and largely non-mathematical way. This chapter includes information on temperature measurement, a very important area of experimental method where thermodynamics is concerned, but it is suggested that a practical study of methods of temperature measurement should take place incidentally to the experiments described later rather than as an independent (and, though important, not particularly interesting) subject in its own right.

Chapter 3 deals with heat transfer and places particular emphasis on the nature of the physical phenomena involved and their relationship to the kinetic theory of heat. The associated experimental work gives many opportunities to gain experience in methods of temperature measurement. It may be remarked that the treatment of heat transfer problems in advance of other aspects of the subject corresponds with the order in which it developed historically; the work of Fourier on heat transfer preceded that of Carnot, Clausius and the other founders of thermodynamic theory.

The following chapters deal with the First and Second Laws of Thermodynamics and other fundamentals. A complete and rigorous treatment of the more difficult aspects of the subject is not attempted since it would be out of place in a book of this nature, while in the experience of the authors the introduction of comprehensive and universal arguments at an early stage tends merely to confuse the student. A comprehensive grasp of the subject is better gained step by step.

In Chapter 10 matters dealt with earlier in the book are reviewed, and where necessary the theoretical treatment previously given is extended and made more precise.

The book closes with an Appendix entitled "Laboratory Practice and Experimental Method".

> The student is urged to study this appendix before starting the laboratory programme and then to reread it at least once at a later stage, when the significance of the points made will become more apparent.

The relationship between the experiments described and the corresponding theoretical concepts is given in the following table.

Table 1.1   Relation between Experimental Programme and Theoretical Instruction

Experiment	Theory
Chapter 2 Experiments with a model illustrating the kinetic theory of gases, familiarization with different methods of temperature measurement.	Mainly non-mathematical treatment of the following concepts: temperature, open and closed systems, thermal equilibrium, non-flow and steady-flow processes, work and heat, the First and Second Laws, the kinetic theory of heat.
Chapter 3 Experiment 1: Thermal boundary layer on a flat plate.	Heat conduction, Fourier's Law, steady and unsteady heat flow, convection, similarity, Nusselt, Prandtl, Grashof Numbers. Analogy between heat transfer and fluid friction. Radiation, the black body, Stefan–Boltzmann Law. Heat transfer at low gas pressures. Laminar, turbulent and thermal boundary layers, boundary layer measurements.
Experiment 2: Heat transfer by forced convection from a single tube and a tube bank.	Exponential law of cooling, presentation of results in terms of Reynolds, Nusselt and Prandtl Numbers.
Experiment 3: Boiling heat transfer.	Evaporation and condensation, nuclear and film boiling, burn-out.
Experiment 4: Heat transfer by radiation, combined heat transfer by free convection and radiation.	Radiation, free convection from horizontal cylinder, influence of pressure, convective heat transfer at low pressures, Knudsen Number. Solar radiation.
Chapter 4 Experiment 5: Energy conversion in a compressed-air motor.	The steady-flow energy equation. Non-flow and steady-flow processes, mechanical work, First Law of Thermodynamics, measurements of power and efficiency, throttling process.
Chapter 5 Experiment 6: The Heat Pump	Second Law of Thermodynamics, reversibility, entropy, entropy dia-

Experiment	Theory
	grams, availability, the heat pump, the Carnot cycle.
Chapter 6 Experiment 7: Performance of a steam power plant.	Thermodynamic properties, equations and diagrams. Latent heat, h–s diagrams for perfect gas, air and water, the Rankine cycle, vapour compression refrigeration cycle.
Chapter 7 Experiment 8: Air-conditioning apparatus.	Properties of mixtures, psychrometric measurements, dew-point.
Chapter 8 Experiment 9: Flow and combustion processes in a gas combustion chamber.	Combustion processes, calorific values, gas analysis, dissociation, properties of flue gases.
Chapter 9 Experiment 10: Performance of a high-speed centrifugal compressor.	Isothermal and isentropic compression, real compression processes, efficiencies, representation of processes on h–s diagram.
Experiment 11: Performance of a small diesel engine.	4-stroke cycle, indicator diagrams, measurement of power, the energy balance, efficiency, air standard cycles.
Experiment 12: Variable compression-ratio petrol engine.	Effect of mixture strength and compression ratio on petrol engine performance, the "hook" curve, the combustion process in a spark ignition engine.

2

Fundamentals; the Measurement of Temperature

2.1 Introduction

It is a matter of everyday experience that some bodies feel hotter than others and that we can arrange them in order of temperature by means of a thermometer, an instrument in which we observe the expansion of a liquid in response to an increase in temperature.

Experience also shows that when we bring two bodies, initially at different temperatures, in contact with one another their temperatures in due course become equal. If we place a piece of hot iron in water the iron is cooled and the temperature of the water increases until they both arrive at the same temperature.

Two bodies are said to have equal temperatures when (in the absence of chemical reaction) no change in their characteristics can be observed when they are brought into contact. They are then said to be in thermal equilibrium. If we bring a thermometer into contact with two bodies in succession and find that the thermometer readings are identical, then the two bodies when brought into contact with each other will show no change of properties. This principle of thermal equilibrium is sometimes called the Zeroth Law of Thermodynamics.

If a number of bodies are in thermal equilibrium then neither the nature of the body nor the process by which thermal equilibrium has been reached, whether by cooling, heating, evaporation or condensation, makes any difference to the final temperature. Temperature is thus a measure of the state of a body or system that is independent of its previous history. Measurements having this character are known as thermodynamic properties.

It is another commonplace that the expenditure of mechanical work in the form of friction can give rise to heat. The "fire-drill", known to primitive peoples in many parts of the world, is an example of a very early application of this principle. It is remarkable that its converse, the transformation of heat into mechanical work, was not developed until so late in the history of civilization. The cannon, an elementary form of internal combustion engine, was known in the twelfth century but it was only in the eighteenth century, with the invention of the atmospheric engine, that the development of means of converting heat into useful mechanical work began. The steam engine had already reached a comparatively advanced stage before the fundamentals of the science of thermodynamics were laid down. The incentive for the development of the science was, indeed, largely the desire to improve the performance of the steam engine and right up to the present day the usual method of presenting the theory of thermodynamics reflects this historical influence.

2.2 The Kinetic Theory of Heat

The kinetic theory of heat seeks to explain the thermodynamic behaviour of substances, particularly liquids and gases, in terms of their molecular structure. The theory is not of great use as a basis for calculating the properties of substances, and few experimental demonstrations are possible. It can, however, play an important role in clarifying and unifying the understanding of thermodynamic phenomena. For instance, pressure and temperature, which appear to be quite different entities, are shown each to be manifestations of molecular movement.

The kinetic theory states that the individual molecules of all substances are in a state of irregular motion. In the case of liquids and gases the relative position of individual molecules is not fixed and the observation that all gases and most liquids are capable of mixing completely and in any proportions confirms this view. If a gas were not an assembly of minute bodies but a continuum then it would be difficult to visualize the possibility of unlimited mixing.

The assumption of molecules that are free to move also explains the lack of specific form of fluids. In the case of solids, on the contrary, the molecules are fixed in location relative to their neighbours though with a greater or less degree of freedom to vibrate about a mean position. We can represent a solid body as equivalent to a number of concentrated masses joined to one another by springs, Fig. 2.2.1, the springs representing the inter-molecular forces. We can describe the loss of energy that takes place on impact in terms of this model. Imagine a steel ball to fall from a height and strike the assembly of masses and springs. Some of the masses will be displaced, the associated springs will be compressed or extended and an elastic wave will be transmitted through the body. In due course the ball will rebound, but not to its original height. Some of the kinetic energy of the ball will have been transmitted to the body in the form of motion of the constituent masses and compression of the springs. After a period, with reflection of the elastic wave back and forth, this energy will be absorbed in the form of irregular movement of the constituent masses. The mechanical energy that has been lost from the falling ball eventually manifests itself as a small increase in the general level of vibration about their mean position of the individual point masses forming the body.

If the body on which the steel ball falls deforms drastically we may describe what occurs, in terms of our model, as equivalent to the breakage of some of the springs

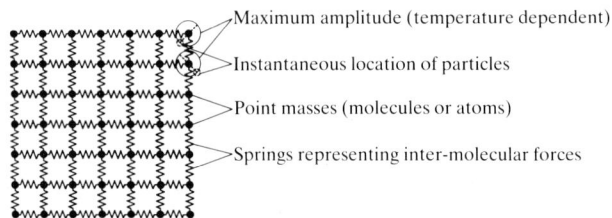

Maximum amplitude (temperature dependent)

Instantaneous location of particles

Point masses (molecules or atoms)

Springs representing inter-molecular forces

Fig. 2.2.1 Model of a solid body

linking the constituent masses with relative sliding and the formation of new linkages. Once more the lost kinetic energy eventually appears in the form of an increased level of vibration of the individual particles.

In both cases, part of the kinetic energy of the falling ball, representing a common motion of its constituent molecules in the same direction, is dissipated in the form of random motions of the molecules of the body upon which the ball falls. This motion is evidently disordered and irregular, but this does not signify that the laws of mechanics have been contravened, only that it is not possible to observe and measure all these individual motions. The only observation accessible to us is the mean level of vibration of the molecules. The absolute temperature of a body is nothing other than a direct measurement of the mean kinetic energy of its constituent molecules.

The kinetic theory of heat leads automatically to the conclusion, confirmed by experience, that the temperature of a body cannot be indefinitely reduced. There exists an absolute zero of temperature at which the kinetic energy of the molecules is zero. There is, however, no upper limit to the temperatures that are possible.

Friction between solid bodies may also be described in terms of the kinetic theory. Even very smooth surfaces display roughnesses of a height equal to many molecular diameters and when such surfaces are pressed together and subjected to relative sliding the surface molecules are forced out of their equilibrium positions and their level of vibration is increased. To bring about this increase in disordered kinetic energy it is necessary to expend mechanical work in maintaining the relative sliding of the bodies. If the force giving rise to the relative sliding is removed, the bodies come to rest and the whole of the ordered energy represented by the sliding force is transformed into the disordered energy of molecular movement.

The process of heat transmission may be described in similar terms. A locally increased level of vibration of the molecules, corresponding to an increased temperature, is transmitted through the body by virtue of the coupling of individual molecules with their neighbours. If in the course of time a uniform level of temperature is developed, this corresponds to an equalization in the mean level of vibration throughout the whole body. It is perhaps easier to visualize this process in terms of a model represented by a series of pendulums, Fig. 2.2.2.

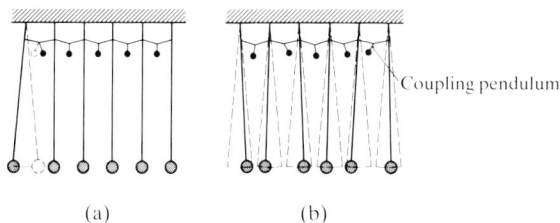

Fig. 2.2.2   A model to represent heat transmission by conduction:
    (a) Left-hand pendulum struck (=local heating)
    (b) After a period all pendulums swing with equal amplitude (=temperature equilibrium)

It will be clear from the above that many thermodynamic phenomena may be described in terms of the laws of mechanics with the additional hypothesis that matter is built up of molecules capable of relative motion and oscillation. An implication is that at the molecular level friction does not occur. At the macroscopic level

7

mechanical energy is dissipated as a result of friction, but in the absence of energy exchange with their surroundings the level of the kinetic energy of the individual molecules of a body remains constant.

The kinetic theory throws light on the mechanism by which changes of state from the solid to the liquid and from the liquid to the gaseous phase occur. As the temperature of a solid increases, the amplitude of vibration of the molecules increases also until a temperature is reached at which the inter-molecular bonding breaks down; the individual molecules become capable of independent motion. A weak force remains which ensures that molecules do not become completely separated from their neighbours, but they are free to travel within the body of the substance which has now become liquid. The oscillatory motion of the molecules associated with the solid state has become an irregular zigzag motion. This motion is also subject to Newton's Laws and to the requirements of quantum mechanics. The kinetic energy of the random translatory motion of the molecules has become a measure of the temperature of the liquid.

Observation shows that during the melting process a substantial amount of heat must be added without any resultant increase in temperature. The kinetic theory interprets this fact in terms of the energy that must be supplied to break the inter-molecular bonds associated with the solid state. This energy does not appear as increased kinetic energy of the molecules and is termed latent heat or, in the terms of the kinetic theory, potential energy. The total energy of a molecule in the liquid state is substantially greater than that in the solid state even though the kinetic energy of both, and hence their temperature, may be equal. A mixture of solid and liquid at the melting point may thus be in thermal equilibrium. On freezing, the latent heat associated with the melting process is released.*

An analogous process, which may also be explained in terms of a molecular theory, takes place during the evaporation of a liquid. In the gaseous state the inter-molecular binding forces are negligible and the individual molecules travel at high velocity, with frequent collisions with other molecules or with the walls of the containing vessel.

Observations of this kind lead us to distinguish between temperature and internal energy. The addition of heat to a substance can result in either an increase in temperature, corresponding to an increase in the kinetic energy of the molecules, or in an increase in the energy stored in other forms. The total energy that must be supplied to a substance to raise its temperature to a given level is termed the internal energy.

The mean velocity c of the molecules of a perfect gas (see Sections 6.2 and 10.5) is given by:

$$c = \sqrt{3RT}$$

where R = Gas Constant
T = absolute temperature

For nitrogen at a temperature of 300K (27°C), $c = 516$ m/s; for hydrogen at the same temperature $c = 1930$ m/s.

The kinetic theory indicates that the mean kinetic energy of translation of a gas

---

* The inter-molecular binding forces referred to above must not be confused with the very much greater forces of attraction and repulsion that exist between the individual atoms.

molecule at a given temperature is equal for all gases despite their different molecular masses:

$$E_{kin} = \tfrac{3}{2}kT$$

where $k = 1.38 \times 10^{-23}$ J/K, the Boltzmann Constant
T = absolute temperature

Thus, lighter molecules must possess higher velocities than heavier ones to give equal kinetic energies. These differences in velocity of translation persist in gaseous mixtures.

That the collisions between gaseous molecules and between the molecules and the walls of the container must be fully elastic is shown by the fact that, in the absence of heat loss, the temperature of the gas, itself a measure of the kinetic energy of the molecules, remains constant.

The question naturally arises as to whether, in addition to energy associated with translation, molecules may also possess kinetic energy associated with rotation. This is the case with molecules consisting of more than one atom. Diatomic molecules may be visualized as having a dumb-bell-like form, and molecules having larger numbers of constituent atoms as forming configurations of concentrated masses in space, Fig. 2.2.3. A monatomic molecule possesses three degrees of freedom (translation in the X, Y and Z directions); a diatomic atom has five degrees of freedom, including rotation about two axes perpendicular to each other and to the axis of the constituent atoms. A molecule having three or more constituent atoms has six degrees of freedom, three associated with translation and three with rotation. The kinetic theory states that a molecule in the gaseous state possesses a kinetic energy $\tfrac{1}{2}kT$ for each degree of freedom and that the energy is equally divided between the different degrees of freedom.

A monatomic molecule thus possesses kinetic energy $\tfrac{3}{2}kT$, a diatomic $\tfrac{5}{2}kT$, a molecule having three or more atoms $\tfrac{6}{2}kT$.

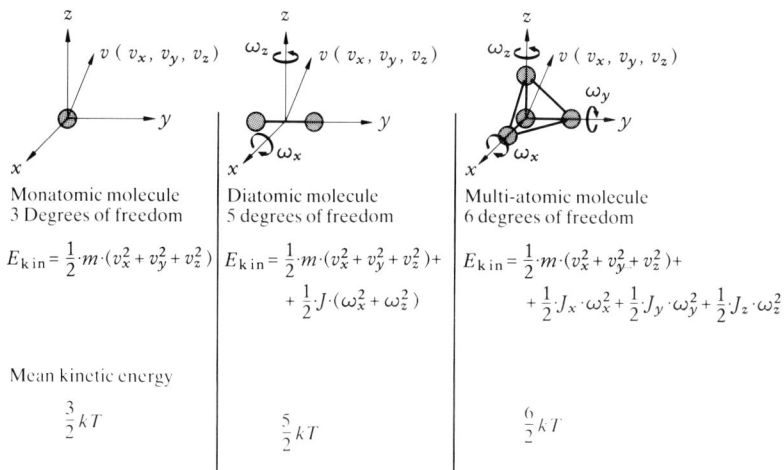

Monatomic molecule
3 Degrees of freedom

$$E_{k\,in} = \frac{1}{2} \cdot m \cdot (v_x^2 + v_y^2 + v_z^2)$$

Diatomic molecule
5 degrees of freedom

$$E_{k\,in} = \frac{1}{2} \cdot m \cdot (v_x^2 + v_y^2 + v_z^2) + \frac{1}{2} \cdot J \cdot (\omega_x^2 + \omega_z^2)$$

Multi-atomic molecule
6 degrees of freedom

$$E_{k\,in} = \frac{1}{2} \cdot m \cdot (v_x^2 + v_y^2 + v_z^2) + \frac{1}{2} \cdot J_x \cdot \omega_x^2 + \frac{1}{2} \cdot J_y \cdot \omega_y^2 + \frac{1}{2} \cdot J_z \cdot \omega_z^2$$

Mean kinetic energy

$$\tfrac{3}{2}kT \qquad \tfrac{5}{2}kT \qquad \tfrac{6}{2}kT$$

Fig. 2.2.3   Kinetic energy of gaseous molecules

9

It is interesting to note that as a consequence of unsymmetrical collisions between molecules on average as much energy of rotation as of translation is transmitted. It is not so easy to understand why a monatomic gas should possess no energy of rotation. One could imagine that such a molecule could be represented by a perfectly smooth sphere which it was impossible to set in rotation, but this does not sound plausible and would be contrary to the indications of atomic physics. The solution to this problem is found in quantum mechanics, an area of theory that is remote from the interests of most engineers. Similar considerations apply to the diatomic molecule which exhibits no rotational energy about the axis joining the constituent atoms.

We have noted above that the mean velocity of gas molecules at room temperature is extremely high. Collisions between the molecules, however, take place very frequently so that each molecule follows an irregular zigzag path (Brownian motion) and its rate of travel in any particular direction tends to be small, Fig. 2.2.4. This is confirmed by observations of the rate of diffusion in gases. The speed with which a perfume diffuses through a large room is very much less than the molecular velocity. A measure of the frequency of collision between molecules is given by the mean free path, or the average distance travelled between collisions. For air at atmospheric conditions this amounts to about 0.6×10^{-4} mm.

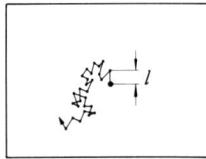

Fig. 2.2.4 Random motion of a gas molecule. Mean free path l=average distance travelled between impacts

The pressure exerted by a gas represents simply the sum of the impulsive forces exerted on the wall of the container as a consequence of the impact of the gas molecules. Application of the force–momentum theorem indicates that this pressure is given by the expression:

$$p = \tfrac{1}{3}\rho c^2$$

where ρ=density of gas.

Heat transfer from a solid to a fluid at lower temperature may be explained in terms of the greater intensity of vibration of the constituent molecules of the solid, corresponding to its higher temperature. As a consequence additional kinetic energy is imparted to the fluid molecules impinging on the solid surface.

The fact that certain chemical reactions result in the release of heat may be explained in terms of the severing of inter-molecular bonds with a resultant increase in the velocity of translation of the constituent molecules, corresponding to a higher temperature, Fig. 2.2.5.

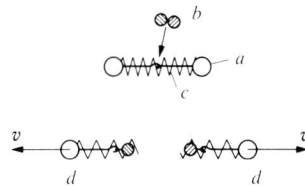

Fig. 2.2.5 Evolution of heat by chemical reaction:
(a) Energetic molecule (e.g. a fuel)
(b) Reactant (e.g. O_2)
(c) Compressed spring representing chemical energy
(d) Products of reaction fly apart; velocity and hence temperature rise

2.3 The Structure of Engineering Thermodynamics: Thermodynamic Systems

It will be apparent that the kinetic theory gives many insights into the mechanism of thermodynamic processes. It is, however, of little assistance in the quantitative treatment of the subject. For this purpose it is more useful to regard solids, liquids and gases as continua and to describe their behaviour in terms of laws and principles that do not refer directly to the kinetic theory and which are based on observations on the macroscopic scale.

The structure of engineering thermodynamics may be represented as follows:

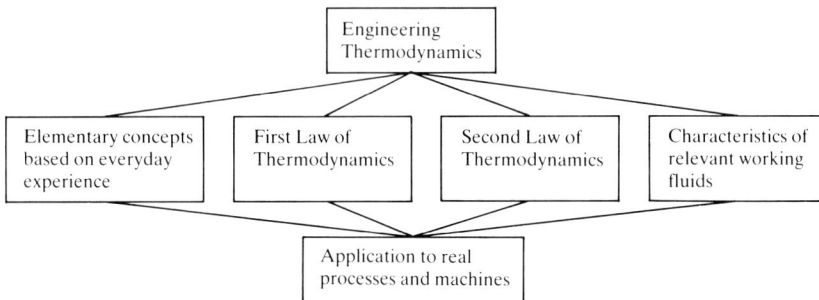

The basic theoretical structure was perfected nearly 200 years after the essential principles of mechanics were understood and more than 100 years after the first successful steam engine had been built. Newton's *Principia* appeared in 1687, but the key contributions to thermodynamic theory appeared in 1822 (Fourier), 1824 (Carnot), 1850 (Clausius) and 1851 (Kelvin).

The probable reason for the late development of thermodynamic theory lies in the relative difficulty of defining a thermodynamic "system". Before any physical phenomenon can be investigated in a scientific manner it is necessary to isolate the phenomenon from its surroundings; a further requirement is for the definition of suitable properties in terms of which the characteristics of the system and of its interaction with the surroundings may be described.

Fig. 2.3.1 shows three systems of increasing complexity: a mechanical system, a

11

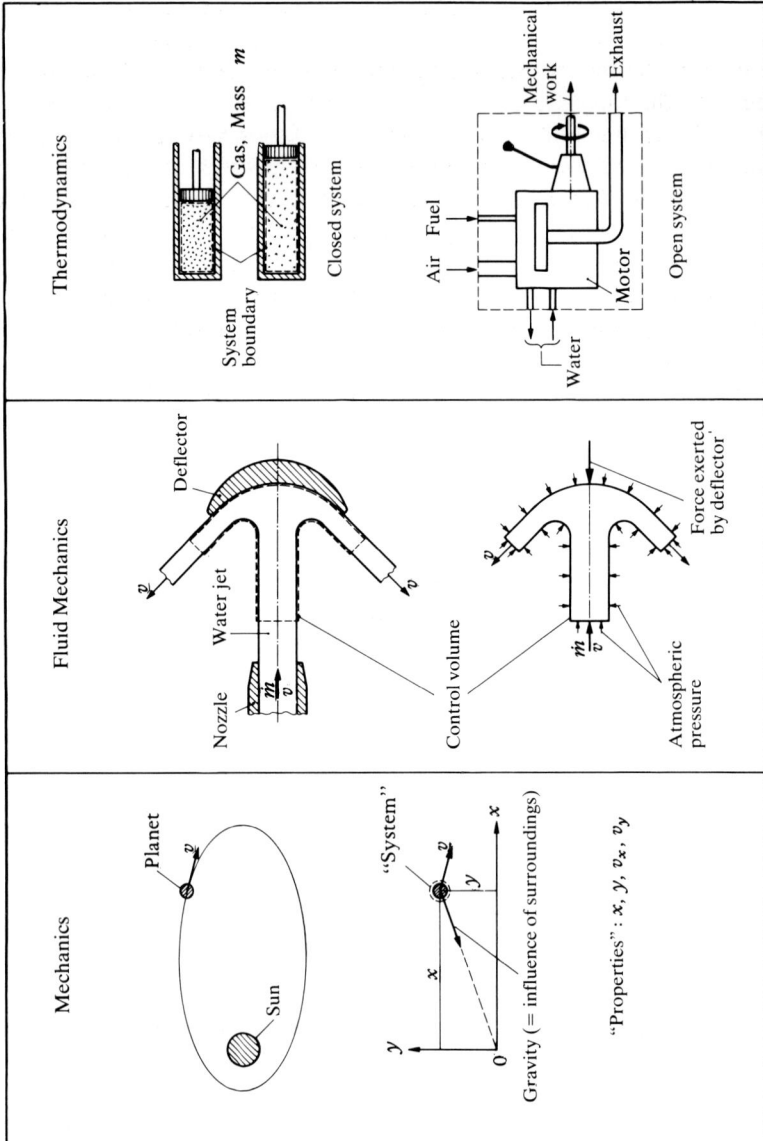

Fig. 2.3.1 Various systems

system associated with fluid flow, and a thermodynamic system. The first example concerns the motion of a planet round a sun. The system may be simply defined as comprising the space containing the planet but nothing else. The properties necessary for a description of the behaviour of the system are also fairly easily defined: they comprise co-ordinates describing the position in space of the planet relative to the sun, vectors describing the motion of the planet, and a further vector representing the force exerted on the planet by the sun, together with the mass of the planet.

In the case of the second example, taken from the field of fluid mechanics, and concerned with the impact of a jet upon a fixed deflector, the system boundary is defined by a dotted line and in the terminology of fluid mechanics is described as a control surface. An analysis of the force exerted on the deflector requires a knowledge of the mass flow rate of discharge from the nozzle and of the velocities and directions of the fluid stream entering and leaving the control surface. This is an example of an open system since mass as well as energy crosses the system boundaries.

Fig. 2.3.1 also shows examples of open and closed thermodynamic systems. The closed system comprises a mass of gas contained in a cylinder. The system boundary is shown as a dotted line enclosing the gas, and the description "closed system" implies only that no mass crosses the boundary of the system; both heat and mechanical work are regarded as able to cross the boundary of a closed system. An adiabatic system is one in which work but not heat is permitted to pass the system boundary. Such a system would be represented by that shown in the figure if it were possible to construct the piston and cylinder of totally non-conducting material; adiabatic systems cannot, of course, be realized completely in practice.

An open thermodynamic system is also shown in Fig. 2.3.1 and represents an internal combustion engine. Here mass flows of fuel, air, cooling water and exhaust gas cross the boundary of the system as well as heat and mechanical work. In general, closed systems are appropriate for the analysis of non-steady-state processes, while

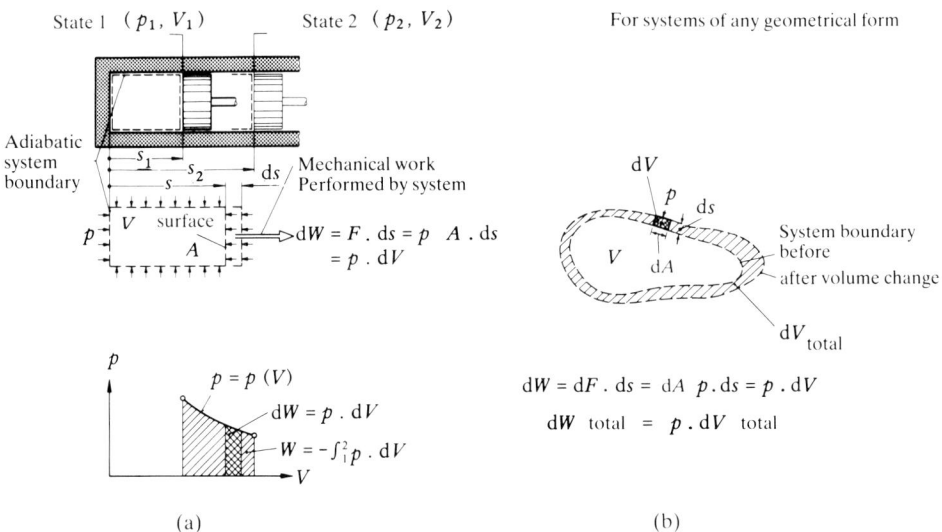

Fig. 2.3.2 Closed adiabatic system:
(a) Piston and cylinder (b) System of any form

13

open systems are appropriate to the steady-flow processes that comprise the vast majority of practical applications.

The choice of suitable properties to describe the state of a thermodynamic system is less simple than is the case with purely mechanical or fluid flow systems, and this difficulty was a further reason for the comparatively slow development of the subject. Temperature is one of the thermodynamic properties and we shall be considering the others later.

Fig. 2.3.2(a) represents a closed adiabatic system and describes the transfer of work across the system boundary to the surroundings in the course of an expansion from state 1 to state 2. Fig. 2.3.2(b) illustrates the generalized form of the same process. In general, system boundaries are capable of conducting heat, and the the thermal balance of the system must take into account heat transfer through the boundaries between the system and its surroundings.

2.4 Thermodynamic Properties: Temperature

In order fully to describe the condition of a body or a system it is necessary to know not only its position, velocity and external shape but its internal condition. Thermodynamics is concerned particularly with this internal condition, by which we understand the total of the measurable properties of the material contained in the system. Experience shows that these properties are not independent one from the other, to the extent that if a small number of the properties are known all the others relating to the system are automatically determined.

Of the properties appropriate to the description of a thermodynamic system three are already known from mechanics and from the mechanics of fluids: mass, volume and pressure (the last two may themselves be defined in terms of mass, length and time). These each comply with the requirement that a thermodynamic property should have a value independent of the process by which the state of the system described by the property was arrived at. Temperature is an additional property required for the description of the state of thermodynamic systems. For the description of kinematic phenomena we require only the properties of length and time, to deal with dynamics we require the additional property of mass, in the field of electricity the property of charge and for thermodynamics the property of temperature. As a fundamental unit, temperature requires no definition.

Temperature differs from the other fundamental properties of length, mass and time in that it loses significance at the molecular level. It is not possible to speak of the temperature of an individual molecule, only of its kinetic energy, which may be described in terms of mass, length and time. Nevertheless in the field of thermodynamics we must accept temperature as a property in its own right.

It is not possible to measure temperature in the simple way in which mass, length and time may be measured, i.e. by determining how many times a unit of standard size is contained in the quantity to be measured. The practical measurement of temperature depends upon the circumstances that, for many pure substances, clearly identifiable changes, such as melting or freezing, are associated with definite temperatures.

The most familiar method of measuring temperature, the mercury thermometer, uses the thermal expansion of mercury as an indication of temperature. The division of the scale of a mercury thermometer so that equal increases in volume are assumed to correspond to equal increases in temperature is more or less arbitrary and depends on the characteristics of mercury. The gas thermometer provides the basis for a temperature scale that is not arbitrary and that also gives an indication that an absolute zero of temperature exists, Fig. 2.4.1. Temperature is measured either as a function of the pressure exerted by the gas (constant-volume gas thermometer) or of the volume of the gas (constant-pressure gas thermometer).

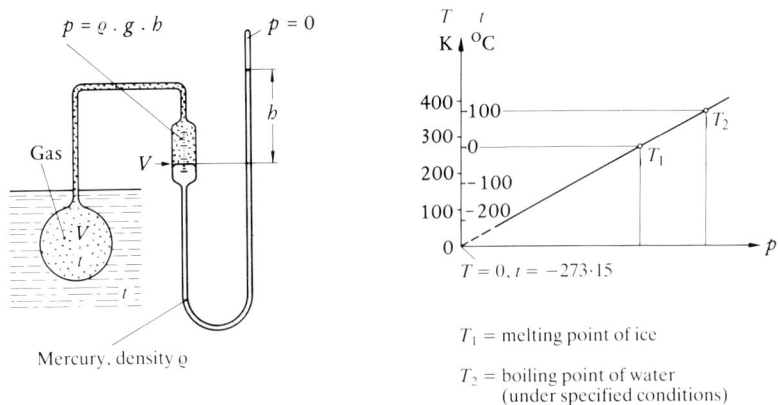

Fig. 2.4.1 Constant-volume gas thermometer

A number of different temperature scales have been used in the past; today only the Celsius and Kelvin or absolute scales are of significance. The relationship between them is given by:

$$T = 273 \cdot 15 + t$$

For most practical applications the constant is rounded to 273. The unit of temperature difference may be described either as the Kelvin or °C.

2.5 The First Law of Thermodynamics

The historical development of the First Law of Thermodynamics may be summarized as follows. In 1822, J. B. J. Fourier published a book with the title *Théorie Analytique de la Chaleur* concerned with the conduction of heat in solids. The partial differential equations developed in this work are still employed. Fourier envisages heat as an imponderable indestructible substance that could both flow in bodies and be stored in them, the latter process being associated with an increase in temperature. Fourier was not concerned with the interchange of mechanical work into heat.

At this stage of the development of thermodynamics heat was regarded, like temperature, as a thermodynamic property. The unit of heat was the calorie, the heat required to increase the temperature of 1 gram of water by 1 deg. C. The relation between the heat supplied to a body and its increase in temperature was given by:

$$Q = mc\Delta t \tag{2.1}$$

where c = specific heat, at that time defined in the unit calories/gram/deg. C.

In connection with the development of the steam engine, efforts were made to establish the relationship between work and heat. In 1842, R. Mayer published a paper in which he stated that heat and mechanical work were equivalent in accordance with the following equation, the historical form of the First Law:

$$Q = JW \tag{2.2}$$

where J, the mechanical equivalent of heat or Joule's equivalent, is not to be confused with the Joule J, the SI unit of work or energy ($= 1$ Newton-metre).

Mayer calculated J on the basis of accepted values for the specific heat of air, and subsequently J. P. Joule made exact measurements using the apparatus shown in Fig. 2.5.1, comprising an insulated calorimeter containing water. Energy delivered by a falling weight was converted into heat by stirring the water. Then:

$$Q = m_w c\Delta t = JMgh \tag{2.3}$$

where m_w = mass of water ($+$ "water equivalent" of calorimeter), $c = 1.0$ for water, by definition, and Mgh represents the work performed by the mass M in falling a distance h.

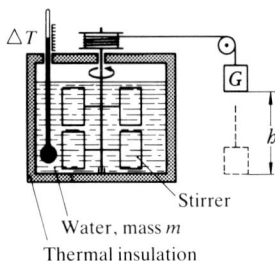

Fig. 2.5.1 Joule's apparatus

Joule's experiment led to a value of the mechanical equivalent of heat which has subsequently been refined and forms the definition of the calorie:

1 kilo-calorie $= 4.1868$ kJ ($= 4186.8$ Nm).

Once it became accepted that heat and work were equivalent entities, it became clear that heat need no longer be regarded as a separate fundamental unit. It is a general principle of science that the number of independent units should be kept to a minimum; given the unit of length, no one would propose a separate new unit for height or for surface area. It therefore becomes desirable to define heat in terms of the First Law.

Before doing so it is necessary to define a further important thermodynamic property: the internal energy. Fig. 2.5.2 represents an adiabatic system comprising a volume of gas contained in a cylinder closed by a piston. The cylinder also contains an electrical resistance element and a stirrer which can be driven from an external source of power. The condition of the system can be changed from initial state 1 to final state

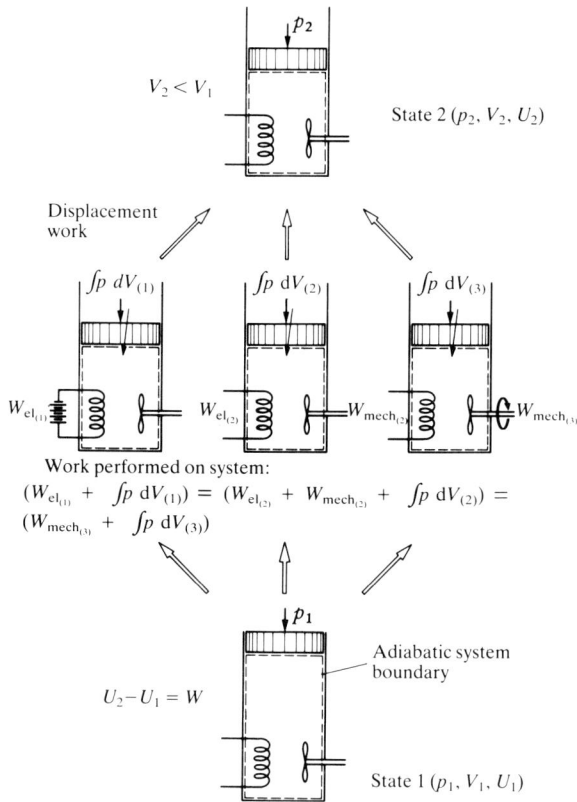

Fig. 2.5.2 Changes in internal energy

2 by the performance of work upon the system, and there are any number of different ways of bringing about this change of state by the supply of work in different proportions of displacement work, electrical work and mechanical work. Experience shows that in order to bring about this change of state only the sum of the work supplied is significant; the route followed by the system from state 1 to state 2 is unimportant. We may define a new property, internal energy:

$$U_2 - U_1 = - W \tag{2.4}$$

In the English-speaking world it is the accepted convention that work performed *by* a system on its surroundings is regarded as positive, while work performed *on* the system is negative, hence the negative signs in equation (2.4) and Fig. 2.5.2. Increase in internal energy, $U_2 - U_1$, equals work input to system.*

Once work has entered the system it is no longer possible to determine whether that work was supplied in mechanical, electrical or displacement form, or as a combination of these forms and whether it was supplied slowly or rapidly. The work stored

---

*On the Continent of Europe this convention has been reversed; work performed on the system is regarded as positive.

in the system as internal energy can be withdrawn and on leaving the system can again take various forms.

If we now assume that, in addition to work, heat may be transferred through the system boundary, equation (2.4) no longer applies and we define the heat supplied by the additional term Q_{12}. (The suffix implies that the work and heat concerned have been supplied in the course of a change of state of the system from state 1 to state 2.)

$$U_2 - U_1 = Q_{12} - W \tag{2.5}$$

Heat supplied *to* the system is regarded as positive in sign while heat leaving the system is regarded as negative.

The First Law of Thermodynamics may now be stated in the following forms:

(a) The increase in internal energy of a closed system is equal to the sum of the heat and work inputs to the system.

(b) Heat supplied to a closed system equals the sum of the work performed by the system plus the increase in its internal energy.

While the English sign convention is not perhaps as logical as the Continental, it has the advantage of leading to statement (b), which clearly applies directly to machines that receive heat and develop mechanical power.

Note that while heat and work are both forms of energy that may be transmitted through the system boundary, resulting in changes in the state of the system in accordance with equation (2.5), they are not, as is internal energy, thermodynamic properties of the system.

In the SI system of units J has ceased to play a part, since both heat and work are measured in terms of the same unit, the joule ($= 1$ Newton-metre). The units in which heat, internal energy and work were previously measured were defined by the First Law in its historical form; the First Law in its present form, equation (2.5), may be regarded rather as a definition of internal energy and heat in terms of work.

Before and indeed since the establishment of the First Law of Thermodynamics innumerable attempts have been made to devise a machine that would produce a net output of work indefinitely. Such a device is known as a perpetual motion machine of the first kind; it cannot, of course, be realized.

The First Law of Thermodynamics has invariably been confirmed by experience, whatever kind of energy and system is involved, and as the Principle of the Conservation of Energy it has achieved a universal significance.

Other formulations of the First Law having particular reference to open systems will be described later.

2.6   The Second Law of Thermodynamics

The most important practical application of thermodynamics is undoubtedly in the conversion of energy in the form of heat into work, either mechanical or electrical. With the onset of a world-wide energy shortage it will be one of the prime responsibilities of the mechanical engineer to ensure that this process is carried out with the maximum possible efficiency.

The First Law of Thermodynamics tells us about the equivalence of the two forms of energy but gives no indication of the limitations on the conversion of energy from one form to the other. The Second Law defines absolute limits to the efficiency of conversion. In particular, it states precisely the proportion of the energy contained in a "hot reservoir", at any given temperature that could ideally be converted into work by a cyclic process. This clearly is a matter of great practical importance since it establishes a standard with which the performance of real machines may be compared.

Like the First Law, the Second is the product of experience and its validity rests on the fact that no process has ever been discovered that runs contrary to its predictions. It does not, however, possess the precision of a mathematical law, nor of the First Law; in the microscopic and atomic region the Second Law of Thermodynamics has no validity. It may only be applied to systems and processes involving a sufficient quantity of matter for the random distribution of energy between the individual molecules of a nominally uniform medium to cease to be of significance. This distinguishes the Second Law from the First, which applies rigorously down to the atomic scale.

Many distinguished scientists have produced statements of the Second Law of Thermodynamics, some of which at first sight seem to bear little relation to one another. Several of the most significant are reproduced in Table 2.6.1; each formulation can be shown to be consistent with the others.

Table 2.6.1 Statements of the Second Law of Thermodynamics

(A)   It is impossible to construct a system which, without receiving work from an external source, will transfer heat from a body at lower temperature to one at higher temperature.

CLAUSIUS

(B)   Whenever one form of energy is converted to another some part of this energy appears as heat.

(C)   It is impossible to construct a perpetual motion machine of the second kind.

OSTWALD

(D)   It is impossible to construct a system which will withdraw heat from a reservoir and convert the whole of this heat into work.

PLANCK

(E)   A scale of temperature, having an absolute zero value, can be defined and is independent of the properties of any particular substance. This scale is identical with the idealized version of the gas temperature scale.

LORD KELVIN

(F)   The expenditure of work W can (but does not necessarily) result in the transfer of heat from a reservoir at lower temperature T_2 to one at higher temperature T_1. The maximum possible amount of heat that can be transferred is given by:

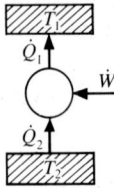

$$\dot{Q}_2 = \dot{W}\frac{T_2}{(T_1 - T_2)}$$

(G)   The maximum possible amount of work W that a system drawing heat from a reservoir at temperature T_1, and rejecting heat to a reservoir at temperature T_2, can produce is given by:

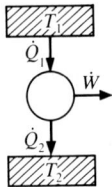

$$\dot{W} = \dot{Q}_1\frac{(T_1 - T_2)}{T_1}$$

Statement A comes closest to everyday experience. Heat can only be transferred from the contents of a refrigerator to the warm surroundings by the expenditure of electrical power; switch off the power and the temperature of the contents eventually reaches that of the surroundings. However complicated we may make it, it has never been possible to construct a system, the overall effect of which has been to transfer heat from a lower to a higher temperature without the expenditure of work.

Statement B is also confirmed by everyday experience. A ball dropped onto a hard surface does not rebound to its original height, the efficiency with which an electrical generator converts mechanical energy into electrical, or that with which an electric motor reverses this process, does not much exceed 90 percent. Whether as a result of impact, mechanical friction or electrical resistance, some part of the energy appears as heat. While the First Law is concerned with the conservation of heat, the Second is concerned with its degradation from forms of "greater" value: mechanical, electrical, heat at high temperature, to forms of "less" value, chiefly heat at low temperature.

A perpetual motion machine of the second kind, Statement C, is one that is capable of drawing heat from a reservoir at some given temperature and converting the whole of that heat into work, without the requirement to reject any heat to some other reservoir at a lower temperature. Such a machine, if it were possible to construct it, would be just as valuable as a perpetual motion machine of the first kind. The whole of the world's energy requirements could be met by lowering the temperature of the oceans by about 1/30 deg. C per year. Inventors, who have for centuries been fascinated by the idea of the perpetual motion machine, are nowadays deterred not so much by the failures of their predecessors as by the fact that Statement C may be convincingly deduced from Statement A, itself evidently in accordance with experience. Statement D is a more scientific expression of Statement C.

The absolute scale of temperature, Statement E, is a fact of experience to the extent that the readings of the gas thermometer, Fig. 2.4.1, approximate more and more closely to this scale as the gas pressure is reduced. This result is the same whatever gas is employed (provided the liquefaction point is not approached). The significance of the absolute temperature scale will emerge later.

With Statements F and G we come to a quantitative expression of the Second Law. The associated diagrams show the conventional representation of a "heat engine" (G) and a "reversed heat engine" (F). Neither statement may be deduced directly from experience, although, among all the millions of prime movers that fall within the category of heat engine, none has exceeded, though some have approached, the efficiency of conversion implied by these statements.

Statement F imputes to work a certain capacity to generate temperature differences. The accompanying diagram represents a "system", a reversed heat engine, which receives a supply of work at rate $\dot{W}$ and acts to withdraw heat at a rate $\dot{Q}_2$ from a reservoir at temperature T_2, and to deliver it, together with a heat flow equivalent to $\dot{W}$ (see the First Law), to one at a higher temperature T_1. It is shown later, Section 5.3, that if the process took place under the most favourable conditions, with no friction or mechanical losses, and with no temperature differences arising as a consequence of heat transfer in and out of the system, then the possible increase in temperature would be given by:

$$T_1 - T_2 = T_2 \frac{\dot{W}}{\dot{Q}_2} \tag{2.6}$$

The implication of this equation, which is a rearrangement of Statement F, is that the quantity of heat that could ideally be transferred from a reservoir of heat to one at a higher temperature varies inversely with the temperature difference, Fig. 2.6.1.

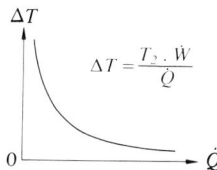

Fig. 2.6.1 Relation between heat transferred and temperature difference

Statement G is evidently the inverse of Statement F. The two, taken together, imply that the amount of work required to transfer a given amount of heat from a reservoir at lower temperature to one at higher is, ideally, exactly equal to the amount of work that could be developed as a consequence of the transfer of the same amount of heat to the low-temperature reservoir from the higher.

The implication is that energy could as well be stored in the form of heat as in the form of work. In practice, however, the process of energy storage in the form of heat invariably involves losses and it is impossible to recover the full quantity of work implied by the equation. The problems involved in the hydraulic "pumped storage" of energy are minor by comparison.

It must be emphasized that both the heat engine of Statement G and the reversed heat engine of Statement F are visualized as operating cyclically; the mass of working fluid within the system circulates continuously, at one stage receiving heat from the hot reservoir, at another rejecting heat to the cold reservoir. Planck was the first to point out that it is possible in a non-cyclic process to achieve 100 percent conversion of heat into work.

Fig. 2.6.2 shows such a process. A gas contained by a piston and cylinder expands, performing work on the piston. Just sufficient heat is added to maintain the temperature of the gas constant. By the First Law, $Q = W$. No means is known by which the gas could be restored to its original condition and the process repeated in the absence of heat rejection *and* with a net output of work.

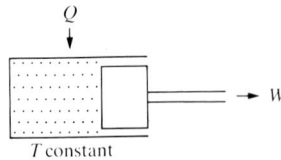

Q

W

T constant

Fig. 2.6.2 A non-cyclic process permitting 100 percent conversion of heat into work

The equation of Statement F may be regarded both as a formulation of the Second Law and as a definition of the thermodynamic temperature scale in terms of work. It may be rewritten:

$$\frac{\dot{W}}{\dot{Q}_2} = \frac{\Delta T}{T_2} = \frac{\Delta t}{f(t_2)} \qquad (2.7)$$

Here T is the temperature on the thermodynamic (Kelvin) scale and t that on some optional scale, e.g. the Celsius scale. For this scale we have already seen that:

$$f(t_2) = t_2; \quad T_2 = t_2 + 273; \quad \Delta T = \Delta t$$

It should be noted particularly that nowhere in this discussion have we referred to the properties of the working fluid used in the system that effects the conversion of work into heat or vice versa. The relations between $\dot{W}$ and $\dot{Q}$ are functions of the reservoir temperatures only. If the use of different working fluids permitted differing (ideal) values of $\dot{W}/\dot{Q}_2$ for given reservoir temperatures then it would in principle be possible to construct a perpetual motion machine of the second kind.

We could arrange a heat engine operating in accordance with Statement G to drive a reversed heat engine, Statement F, and select the working fluids so that, with a work input $\dot{W}$, the reversed heat engine transferred more heat from the cold reservoir to the hot than was rejected to the cold reservoir by the driving engine. Such a system would contravene Statements A, B and C.

Statement G is of particular importance in that it defines the ideal (Carnot cycle) thermal efficiency of a heat engine:

$$\eta = \frac{\dot{W}}{\dot{Q}_1} = \frac{T_1 - T_2}{T_1} \qquad (2.8)$$

The Carnot cycle is described, and the derivation of this equation given, in Section 5.3. The form of the equation implies that the higher the temperature of the hot reservoir and the lower that of the cold reservoir or "sink", the greater the ideal (and hence the potential) efficiency of the heat engine. In the vast majority of practical applications the sink is represented either by the atmosphere or by a mass of water, a river or the ocean.

It is inconvenient that in practice the temperature T_2 of the sink cannot be much less than about 300K. It follows that, for reasonable efficiencies, T_1 must be high and we encounter metallurgical difficulties, for instance in the selection of suitable high-temperature steels for boilers.

While a rigorous derivation of the Second Law is by no means easy, its most important practical corollary, the definition of the maximum possible efficiency of energy conversion implied by equation (2.8), is apparently so simple as to be almost self-evident. Consider two heat engines, one utilizing a source of heat at temperature 400K, and one a source at 600K. Let us assume that in both cases heat is rejected to a sink at a temperature of 300K.

Let us further assume that the relative rates of heat input to the engines are 400 watts and 600 watts. We could then represent their performance by a diagram such as Fig. 2.6.3, in which the ordinate is absolute temperature and the abscissa is some, as yet undefined, unit such that areas in the diagram represent flows of heat or work.

In Fig. 2.6.3(a), area *abef* represents the rate of heat rejection to the sink, 300W, *bcde* represents the output of work, 100W, while the whole area of *acdf* represents the rate of heat input, 400W. In Fig. 2.6.3(b), area *abef*, the heat rejected, is again 300W, but area *bcde*, the output of work, is 300W, while the heat input is 600W.

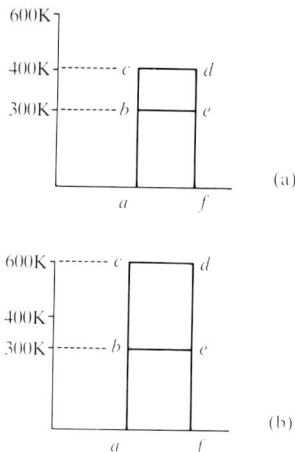

Fig. 2.6.3   Elementary representation of a heat-engine cycle

The corresponding thermal efficiences, equation (2.8), are respectively 100/ 400 = 0·25 and 300/600 = 0·50.

We shall see later that the simple diagrams of Fig. 2.6.3 are closely paralleled by figures in which the vertical axis is represented, as here, by the absolute temperature, while the horizontal axis is formed by a new quantity, entropy. On these diagrams it is possible to represent both theoretical processes and those realizable in practice.

The concept of availability, which plays such a large part in modern thermo-dynamics, and to which we return in Chapter 5, may be introduced at this point. In the case of the heat source at 400K 25 percent of its contents is in theory *available* for conversion into work; for the source at 600K the availability is 50 percent. We see that availability is a function not only of the temperature of the hot reservoir but of the associated sink.

It should be noted that work, unlike heat, is not associated with any particular level of temperature. Thus, for example, we can produce extremely high temperatures by passing electrical power through a fine wire.

Once again the kinetic theory is enlightening. Imagine one side of the body depicted in Fig. 2.2.1 to be subjected to a high temperature. The molecules on that side of the body will vibrate with increased vigour and the Second Law indicates that this increased level of vibration will in the course of time dissipate itself throughout the body, so that all the molecules come to vibrate with the same mean energy. It is plainly entirely improbable that the reverse process should take place, with a transfer of energy from the molecules vibrating with less vigour to those at a higher energy level.

We have seen that whenever energy is transformed from one form to another part of it is dissipated as heat. Real processes always involve friction, impact or loss of energy due to electrical resistance. This heat can never be fully recovered as work. The Second Law of Thermodynamics may be expressed in molecular terms somewhat as follows:

Any thermodynamic system tends always to move from a state of order (low probability) to one of disorder (higher probability).

As an example, consider the impact between two inelastic bodies. Before the impact the component molecules have, superimposed upon their usual random motion, an ordered motion represented by the relative velocity of the bodies. After the impact this motion has been dissipated in the form of disordered molecular vibration.

The state of "order" in a body in which temperature differences exist is greater than in a body at uniform temperature since the intensity of molecular motion is then "ordered" at different levels in different parts of the body.

The concept of ordered and disordered molecular states may be illustrated in terms of the mixing of gases. Fig. 2.6.4(a) represents a vessel separated into two compart-

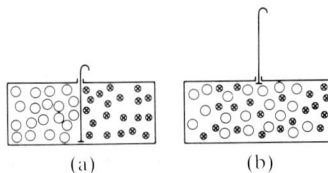

(a)   (b)

Fig. 2.6.4 Mixing of gases:
(a) Vessel containing two gases separated by a partition   (b) Mixing by diffusion

ments, one containing hydrogen and the other nitrogen, both at the same temperature and pressure. If we remove the dividing wall, Fig. 2.6.4(b), the gases will mix by diffusion: order (two kinds of gas, molecules separated) becomes disorder (mixture of the two gases). Both the First and the Second Laws apply to this process: neither a change in temperature nor an output of work results.

2.7 A Model Analogy of the Kinetic Theory of Heat

Fig. 2.7.1 shows a model with the aid of which a number of features of the kinetic theory of heat may be illustrated. It consists of a rectangular glass vessel with a movable floor. This floor may be oscillated in a vertical direction over a range of frequencies. The vessel is partially filled with a large number of steel balls of diameter 2·5 mm, each intended to represent one molecule. As the frequency of oscillation of the vessel floor is gradually increased the balls execute a range of motions analogous to the molecular motions associated with the solid, liquid, vapour and gaseous phases.

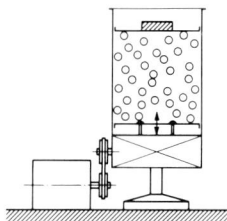

Fig. 2.7.1   Model illustrating the kinetic theory of heat

At the upper end of the frequency range all the balls are in continuous flight and, if we close the vessel by means of a weighted piston it may be demonstrated that the volume occupied by the flying balls varies approximately inversely with the load on the piston, an analogy to Boyle's Law. The analogy is, of course, far from perfect; energy is continuously dissipated and must be replaced by the driving motor, while the force of gravity exercises an effect that is negligible in the molecular situation.

Other phenomena that may be demonstrated with the apparatus include the Brownian motion (with the aid of a single larger ball of less density), diffusion (with the aid of balls of different colours), osmosis (with the aid of a perforated dividing wall), distillation, and hydrostatic pressure.

While formal experiments are inappropriate, it is recommended that the student should observe the various demonstrations after reading the present chapter.

2.8 The Technique of Temperature Measurement

British Standard Code 1041:1943 [12] gives comprehensive guidance on methods of measuring temperature. It should be consulted whenever the reader is confronted with a practical problem of temperature measurement.

The accurate measurement of temperature presents some of the most difficult problems encountered in the field of instrumentation. There are numerous different possible sources of error and these errors can on occasion be very large. In some cases, e.g. during combustion, it may be difficult even to define the temperature one is attempting to measure. A sound grasp of measurement techniques is essential to the engineer and since the techniques of temperature measurement are particularly complex and diverse a proper understanding of them is of considerable educational value.

2.8.1 The International Practical Temperature Scale of 1968 (IPTS–68)

The basic fixed points of the international temperature scale were agreed at the Thirteenth General Conference of the International Committee on Weights and Measures and are given in Table 2.8.1. These refer to a standard atmospheric pressure of 760 mmHg and the code of practice defines this pressure precisely and also gives formulae defining correction factors for other pressures.

Table 2.8.1   The International Practical Temperature Scale of 1968: Fixed Points

Boiling point of oxygen	$-182 \cdot 962°C$
Triple point of water*	$+0 \cdot 01°C$
Steam point	$100°C$
Melting point of zinc	$419 \cdot 58°C$
Melting point of silver	$961 \cdot 93°C$
Melting point of gold	$1064 \cdot 43°C$

* See Chapter 6, Section 6.1.

A number of secondary fixed points have also been specified and the method of interpolating temperatures between the fixed points is also laid down. This involves the use of a platinum resistance thermometer for temperatures of up to 660°C, a platinum–rhodium thermocouple from 660°C to the gold point, and above the gold point a radiometer of specified characteristics. A comprehensive account of the evolution of the International Scale will be found in [15].

2.8.2 Types of Temperature Measuring Instrument

Table 2.8.2 summarizes the principal methods of temperature measurement and indicates the range of temperature and type of application for which they are suitable.

Table 2.8.2   Temperature Measuring Instruments: Ranges and Fields of Application

Type of Instrument	Temperature Range	Type of Measurement
Liquid-in-glass	$-200°C$ to $500°C$	Solids, liquids, gases
Bimetallic	$-30°C$ to $400°C$	Gases
Vapour pressure	$-20°C$ to $350°C$	Solids, liquids, gases

26

Liquid-in-steel	$-35°C$ to $600°C$	Solids, liquids, gases
Electrical resistance	$-240°C$ to $600°C$	Solids, liquids, gases
Base metal thermocouples	$-200°C$ to $1100°C$	Solids, liquids, gases
Rare metal thermocouples	$-200°C$ to $1440°C$	Solids, liquids, gases
Total intensity radiation pyrometer	$-40°C$ upwards	Radiating surfaces
Optical pyrometer	$700°C$ upwards	Radiating surfaces
Light-sensitive cell pyrometer	$700°C$ upwards	Radiating surfaces
Colour paints and pencils	$40°C$ to $1350°C$	Hot surfaces
Fusible indicators	$100°C$ to $1600°C$	Solids
Refractory cones	$600°C$ to $2000°C$	Furnaces and kilns

The methods of temperature measurement in Table 2.8.2 fall into three groups: contact methods, non-contact methods and special methods. Contact methods are used for the great majority of engineering purposes; non-contact methods are appropriate when the temperatures involved are so high that visible radiation is present, while the special methods are used in specialized research situations or for monitoring performance in the process industries.

In contact methods the testing body, e.g. the bulb of the thermometer, is brought into contact with the hot body whose temperature is to be measured: the Zeroth Law of Thermodynamics, p. 5, is applied. It is by no means so easy as it may at first appear to ensure that the temperature of the testing body is equal to that of the body whose temperature is to be measured. The testing body must have a small enough heat capacity not to disturb the temperature conditions and to follow fluctuations of temperature without too much lag. The temperature in the interior of a solid can usually be measured by drilling a hole, inserting the testing body and filling the surrounding space with a suitable liquid. The presence of the hole may itself disturb the temperature distribution in the solid.

There is usually no particular difficulty in measuring mean liquid temperatures, provided the liquid is adequately stirred. Serious difficulties may, however, arise in the measurement of gas temperature. A sensing element immersed in a gas flowing in a pipe takes up a temperature intermediate between that of the gas and that of the pipe surface. The extent of the error depends on the velocity of flow of the gas, the size of the sensing element and the emissivity of the element and of the pipe walls. Errors may be reduced by surrounding the sensing element with a series of concentric screens, Fig. 2.8.1, and for specialized applications, by the use of a suction pyrometer, Fig. 2.8.2. The concentric screens take up a temperature intermediate between that of the gas stream and the walls, thus reducing radiation errors. In the suction pyrometer the sensing element is located in a side passage out of which a sample of gas may be drawn by means of a small vacuum pump. As the gas velocity increases the indicated temperature rises, and a curve of gas velocity against temperature may be plotted. It is assumed that when further increase in velocity results in no change in indication the sensing element is showing the true gas temperature. Experiment 9 involves the use of a suction pyrometer. Care must be taken to avoid condensation of vapour on the sensing element, since the latent heat of condensation or of subsequent evaporation may give rise to serious errors in the reading.

27

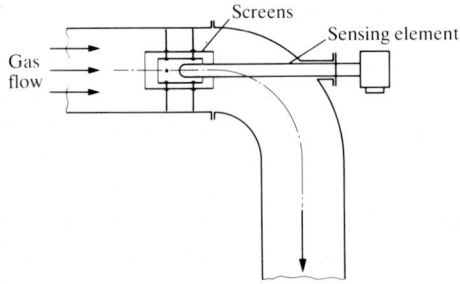

Fig. 2.8.1 Shielded thermocouple for measurement of temperature in a gas flow

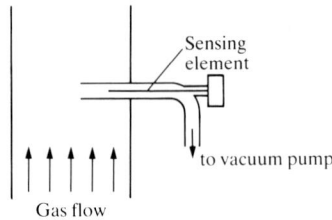

Fig. 2.8.2 Suction pyrometer

The thermocouple is a particularly attractive method of temperature measurement; it can be of small size, it is robust, it lends itself to remote indication and it is capable of following rapid temperature fluctuations more rapidly than any other means of temperature measurement. Certain special precautions are, however, necessary and these will be dealt with later.

Non-contact instruments measure the amount of radiation received from the body whose temperature is to be determined. They will only give exact results if the body is perfectly black (see Section 3.3) and the intervening space perfectly transparent. In practice accurate readings can only be obtained if the body is placed in an enclosure that is at approximately the same temperature as itself, or if the surface has a known emissivity and the temperature of the surroundings is known. In the latter case a correction can be applied. The presence of CO_2, water vapour, smoke or dirty glass between the hot body and the pyrometer will introduce errors which can be large.

Where intense thermal activity is present, as in a furnace, reliable temperature measurements require a specialized understanding of the physics and chemistry of the processes taking place. As we have seen, the temperature of a system is proportional to the molecular energy distributed in the form of translation, vibration and other degrees of freedom. When combustion is taking place it cannot be assumed that normal equilibrium conditions apply or even that any true value can be assigned to the temperature.

Each method of temperature measurement listed in Table 2.8.2 has its own appropriate area of application and presents its own problems. The Code of Practice [12] gives detailed guidelines to assist in making the correct choice; we shall consider two only of the most important methods: liquid-in-glass thermometers and thermocouples.

28

2.8.3 Liquid-in-Glass Thermometers

The familiar liquid-expansion thermometer has many advantages. It is cheap, simple, easily portable and requires no additional indicating instrument. It is also capable of a higher degree of both accuracy and precision than other forms of temperature measuring instrument. For example, the National Physical Laboratory Class A Certificate lays down that a liquid-in-glass thermometer for use within the range $-10°C$ to $+50°C$, with the scale divided in steps of 0.01 deg. C, should have an accuracy within the range $±0.005$ deg. C. This represents a far finer resolution and higher degree of accuracy than is attainable by any other method of temperature measurement. The characteristics of liquid-in-glass thermometers are specified in a number of British Standards [13]. The quality of the glass, its thermal stability and the accuracy of the bore are all critical and the National Physical Laboratory has laid down certain approved thermometric glasses.

The main disadvantages of the simple liquid-in-glass thermometer are that it is rather difficult to read, it has a fairly high thermal capacity and slow response to changes in temperature, and it is, of course, fragile. It is subject to stem errors which arise because liquid-in-glass thermometers are normally calibrated with the complete instrument at the test temperature, while in most applications only the bulb and the lower part of the stem are at this temperature with the remainder of the column exposed to the atmosphere. However, corrections may be applied and are described in the Code of Practice.

Where the liquid-in-glass thermometer cannot be directly immersed in the fluid or gas the temperature of which is to be measured, considerable errors may be introduced, principally due to the conduction of heat by the sheath surrounding the thermometer bulb, as a result of which the thermometric liquid never reaches thermal equilibrium with the surrounding fluid.

For many industrial purposes, however, the main requirement is that thermometers used for routine temperature measurements should give consistent readings, and systematic errors of this kind may not be important.

2.8.4 Thermocouples

While not capable of the precision of the liquid-in-glass thermometer, the thermocouple is a very effective and flexible method of temperature measurement. The majority of industrial temperature measurements are made with thermocouples, frequently in association with instruments giving a digital read-out of temperature.

Temperature measurement with the thermocouple depends on the thermoelectric effect. If two conductors of different metals or alloys are joined at one end and connected to a millivoltmeter at the other, a difference in temperature between the junction and the ends of the conductors connected to the millivoltmeter will give rise to an electromotive force, the magnitude of which is dependent upon the temperature difference.

To be precise, the thermoelectric potential is a function of the temperature difference between the hot junction and the cold junction, Fig. 2.8.3(a). In the case of Fig. 2.8.3(b) the cold junction may be assumed as being at the terminals of the measuring instrument; for precise laboratory use it is usual to immerse the cold junction in melting ice in a dewar flask, thus comparing the measured temperature

29

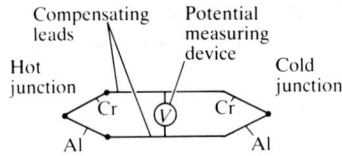

(a)

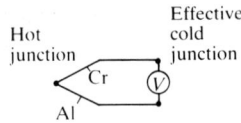

(b)

Fig. 2.8.3   Principle of the thermocouple

directly with one of the fixed points of the International Scale, Table 2.8.1. Industrial thermocouple instruments incorporate a self-compensating cold junction. This is at ambient temperature and adjustment to changes in this temperature takes place automatically.

The thermoelectric effect is very small; typically, a temperature difference of 100°C gives rise to a thermoelectric potential in the range 4 to 5 mV for base metal thermocouples and about 0·6 mV for rare metal couples. Temperature differences at other points in the circuit have no effect on the e.m.f. developed but it is essential that the leads should be identical with the materials of the thermocouple themselves, or should have approximately identical characteristics. Various thermocouple materials have been adopted as standard and their characteristics have been specified [14]. Some of the more important are listed in Table 2.8.3 while, by way of example, values of thermocouple e.m.f. for copper–constantan are shown in Table 2.8.4. The relations between thermoelectric potential and temperature difference are shown in Fig. 2.8.4. Fig. 2.8.5 shows the construction of a typical commercial thermocouple temperature-measuring element.

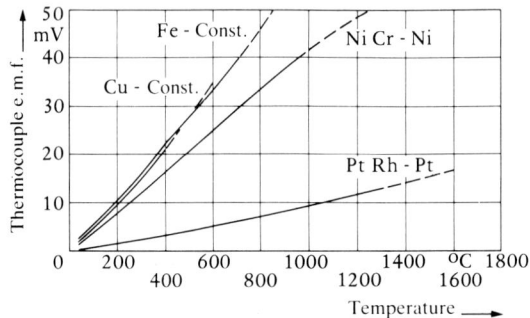

Fig. 2.8.4   Relation between thermocouple e.m.f. and temperature

Table 2.8.3 Principal Thermocouple Materials

Materials	Maximum temperature
Copper(+) Constantan(−)	400°C
Iron(+) Constantan(−)	700°C
Chromel(+) Alumel(−)	1000°C
Platinum–Rhodium(+) Platinum(−)	1400°C

Table 2.8.4 Temperature/e.m.f. for Copper–Constantan (Copper/Copper–Nickel) Thermocouples

°C	0	10	20	30	40	50	60	70	80	90	100
mV	0	0.391	0.789	1.196	1.611	2.035	2.467	2.908	3.357	3.813	4.277
°C	100	110	120	130	140	150	160	170	180	190	200
mV	4.277	4.749	5.227	5.712	6.204	6.702	7.207	7.718	8.235	8.757	9.286
°C	200	210	220	230	240	250	260	270	280	290	300
mV	9.286	9.820	10.360	10.905	11.456	12.011	12.572	13.137	13.707	14.281	14.860
°C	300	310	320	330	340	350	360	370	380	390	400
mV	14.860	15.443	16.030	16.621	17.217	17.816	18.420	19.027	19.638	20.252	20.869

Reference junction at 0°C.

Extracts from BS 4937: Part 5: 1974 are reproduced by permission of the British Standards Institution. Complete copies may be obtained from BSI at Linford Wood, Milton Keynes, MK14 6LE.

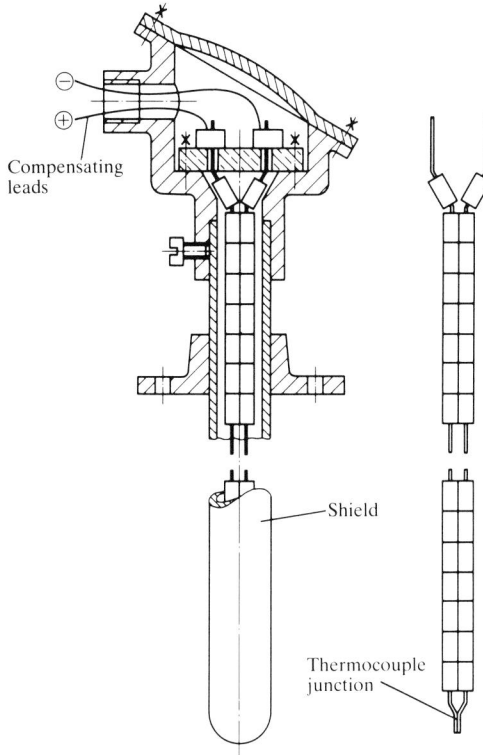

Fig. 2.8.5 Typical industrial thermocouple for measurement of fluid temperature

The compensating leads used to connect the thermocouple to its associated measuring instrument are commonly made of alloys that are less expensive than the thermocouple conductors themselves, but which have thermoelectric characteristics sufficiently close to those of the thermocouple materials with which they are associated to avoid significant error. Compensating leads are colour coded and care must be taken to connect them with the right polarity. Ref. [15] gives a detailed discussion of the technique of temperature measurement by thermocouples.

An important characteristic of a temperature measuring device is the rate at which it is capable of following a temperature change. Fig. 2.8.6 shows the response to a step change in temperature such as would result, for example, from plunging the sensing device into a hot liquid. The form of the curve is given by:

$$T = T_0 + \Delta T(1 - e^{-t/\tau}), \qquad (2.9)$$

where τ = time constant. The physical significance of the time constant is that, if the temperature reading continued to increase at its initial rate the error ΔT would be eliminated in time τ. It is easy to show that one-half ΔT is eliminated after a time $t_{0.5} = 0.693\tau$.

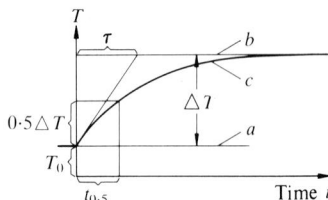

Fig. 2.8.6 Response of temperature-sensing element to a step change:
(a) Initial temperature of element
(b) Temperature of medium
(c) Course of temperature indication

Miniature shielded thermocouples having outside diameters in the range from 2 mm to 0·25 mm are available with time constants down to as little as 0·1 s for gases and 0·01 s when immersed in liquids. Elements such as that illustrated in Fig. 2.8.5 may have time-constants as much as one thousand times as great. An application of miniature thermocouples to the measurement of rapidly varying temperatures is described in Experiment 1.

For most industrial applications, thermocouples are used in conjunction with either an analogue or a digital indicator which is essentially a high-resistance, high-grade millivolt meter calibrated in degrees C. The instrument is designed for use with a particular pair of thermocouple materials and for a specified external resistance. While the thermocouple current is extremely small, there is a finite voltage drop in the compensating leads, and if the resistance departs from the specified value an error is introduced.

This difficulty is overcome if a potentiometer is used to measure the voltage developed by the thermocouple directly. There is then no flow of current in the thermocouple circuit and the potentiometer reads precisely the e.m.f. developed by the thermocouple. For laboratory purposes, a simple hand-adjusted potentiometer with a null-reading galvanometer is suitable; alternatively a self-balancing instrument, in which a restoring mechanism automatically sets the potentiometer galvanometer to zero and gives an automatic indication of the e.m.f., may be used.

A single thermocouple indicator, whether of the direct reading or potentiometer form, is frequently associated with a number of thermocouples, each of which may be coupled to the instrument in turn by way of a change-over switch. Thermocouple switches must be of high quality, with contacts of incorrodible metal if errors arising from contact resistance are to be avoided.

In the control of industrial processes potentiometer-type instruments are frequently associated with pen recorders which give a continuous record of the temperature indicated by a single thermocouple or by a number of couples. In the latter case the instrument is designed to survey and record the indications of all the associated couples at regular intervals.

2.8.5 Exercises in Temperature Measurement

It is a useful exercise to compare the performance and accuracy of the various temperature-measuring instruments available in any particular laboratory.

(a) Measurement of the temperature of a boiling liquid with various instruments and a critical comparison of the results.

(b) Measurement of time constant τ for various instruments, fluids and flow conditions.

(c) Estimation of accuracy of different temperature-measuring instruments.

(d) Calibration of a suction pyrometer.

3

Heat Transfer

3.1 Brief Summary of Theory: Conduction

Conduction is the mechanism responsible for the transfer of heat in solids; it also plays some part in the transport of heat in fluids. The simplest case concerns steady-

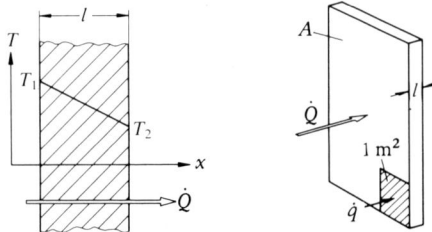

Fig. 3.1.1   Heat conduction through a wall

state heat transfer through a parallel wall, the surfaces of which are at different temperatures, Fig. 3.1.1. Experience shows that the temperature gradient is linear and is related to the heat flow as follows:

$$\dot{Q} = A\frac{(T_1 - T_2)}{l}\lambda, \quad \text{or} \quad \frac{\dot{Q}}{A} = \dot{q} = \lambda\frac{(T_1 - T_2)}{l} \tag{3.1}$$

This is Fourier's Theorem. The thermal conductivity λ is a property of the material; its dimensions are J/ms K. The thermal conductivity of pure copper $= 395$ J/ms K, of steel $= 52$ J/ms K, and of PVC $= 0.12$ J/ms K; the variations between different materials are very wide. The rate of heat transfer per unit wall area $\dot{q}$ is proportional to the temperature difference and inversely proportional to the wall thickness. Observation confirms that under steady-state conditions $\dot{q}$ has the same value at entry and exit; no heat is "lost" as a result of the conductive process.

If heat is transferred through two walls of differing conductivities as in Fig. 3.1.2, $\dot{q}$ is given by:

$$\dot{q} = \lambda_1\frac{\Delta T_1}{l_1} = \lambda_2\frac{\Delta T_2}{l_2} \tag{3.2}$$

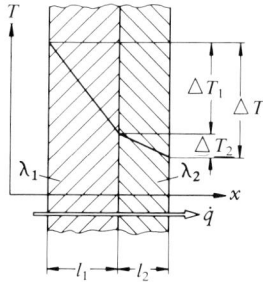

Fig. 3.1.2 Conduction through two walls of differing conductivity in series

The total temperature difference ΔT is then:

$$\Delta T = \Delta T_1 + \Delta T_2 = \dot{q}\left(\frac{l_1}{\lambda_1} + \frac{l_2}{\lambda_2}\right) \tag{3.3}$$

If both walls are of the same thickness that having the poorer conductivity shows the greater temperature fall. The equation assumes that the two walls are in complete thermal contact and this is difficult to achieve in practice. Even the smallest air-space offers a relatively large resistance to the flow of heat.

Thin curved walls, such as tubes having an external diameter not greater than $1 \cdot 5 \times$ internal diameter, may be treated as walls having an area corresponding to the mean diameter of the tube without significant error.

When the conditions are unsteady, for instance during a warming up process, the heat entering the wall is not equal to the heat leaving it; some part is stored in the wall. An energy balance exists, Fig. 3.1.3 and we may write

$$\frac{\Delta Q}{\Delta t} = mc\frac{\Delta T}{\Delta t}, \quad \text{or} \quad \dot{Q} = mc\dot{T} \tag{3.4}$$

where $t =$ time.

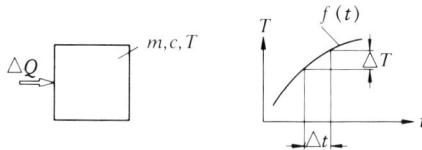

Fig. 3.1.3 Energy balance: non-steady-state heat transfer

This equation forms the basis for the calculation of the results of Experiments 2 and 3.

It necessarily follows that, for finite rates of heat transmission to or from a body of limited conductivity (any real body) the temperature distribution in the body is not uniform. The mathematical theory of heat conduction is largely concerned with the application of equation (3.4) to the calculation of the temperature distribution in bodies of different forms under both steady-state and transient conditions.

For practical engineering purposes it is frequently not necessary to calculate the temperature distribution in detail; a knowledge of the maximum temperature difference associated with a heating or cooling process is sufficient. It may also be adequate to approximate a body of complicated form by a simpler idealized shape.

For flat plates, cylinders and spheres subjected to a linear rate of temperature change at the surface, the temperature profile in the interior of the body assumes a parabolic form after a period of transition following the onset of the temperature change, Fig. 3.1.4(a). The maximum temperature difference ΔT_{max}, applicable either to heating or to cooling, may be taken from Table 3.1.1.

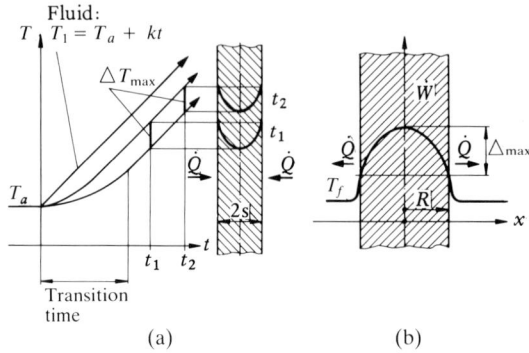

Fig. 3.1.4 (a) Warming of flat plate immersed in fluid of uniformly increasing temperature
(b) Steady-state temperature distribution in cylindrical element with internal heat production (e.g. nuclear reactor fuel element)

Table 3.1.1 Maximum Temperature Difference ΔT_{max} in Bodies of Simple Form

	Flat plate	Cylinder	Sphere
Constant rate of heating or cooling	$\dfrac{l^2}{2a}\cdot\dfrac{\Delta T}{\Delta t}$	$\dfrac{R^2}{4a}\cdot\dfrac{\Delta T}{\Delta t}$	$\dfrac{R^2}{6a}\cdot\dfrac{\Delta T}{\Delta t}$
Uniform evolution of heat within body	$\dfrac{\dot{Q}l^2}{2\lambda}$	$\dfrac{\dot{Q}R^2}{4\lambda}$	$\dfrac{\dot{Q}R^2}{6\lambda}$

$2l$ = thickness of flat plate
R = radius
a = thermal diffusivity
λ = thermal conductivity
$\dfrac{\Delta T}{\Delta t}$ = rate of rise or fall of external temperature

$\dot{Q}$ = rate of heat evolution, W/m³

In calculations dealing with thermal conductivity the thermal diffusivity a, a property of the material, plays an important part. It is defined as

$$a = \frac{\lambda}{\rho c}$$

36

where λ = thermal conductivity, ρ = density of the material and c = specific heat.* It will be apparent that the denominator of this expression represents the thermal capacity of the material per unit volume and that a is a measure of the rate at which a temperature change will be propagated through the material.

Equation (3.4) deals only with heat exchange between a body and its surroundings and the storage of energy within the body. Situations can also arise in which heat is generated within a body, the most important practical case being the generation of heat within the elements of an atomic reactor. In this case also the theory predicts a parabolic temperature distribution, Fig. 3.1.4(b).

3.2 Convection

In the course of the following discussion of free and forced convection frequent reference is necessarily made to aspects of fluid mechanics and in particular to the boundary layer. The discussion will only be clear to the reader who has a working knowledge of these matters. An introductory discussion of boundary layer theory appropriate for this purpose has been given by the authors in [11], Chapter 5.

Consider a vertical wall in contact with a fluid to which the wall is transferring heat, Fig. 3.2.1. We may at first assume that heat transmission in the fluid will be the result

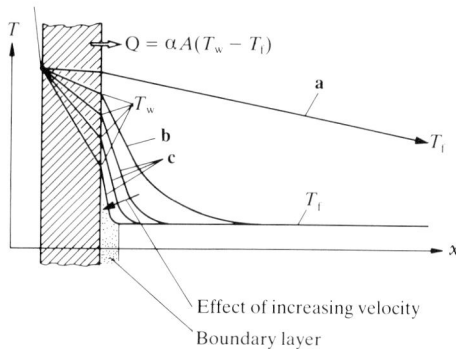

Fig. **3.2.1**   Heat transfer from a surface:
   a   Pure conduction   b   Free convection   c   Forced convection

of conduction, but experience shows that the fluid does not remain at rest in these circumstances; convection currents develop and greatly increase the rate of heat transfer from the wall to the fluid. This process is known as free or natural convection; forced convection occurs when the fluid is also set in motion by other means: when, for example, the wall forms the boundary of a tube through which the fluid is being forced. We may then write:

$$q = \frac{\dot{Q}}{A} = \alpha(T_w - T_f) \tag{3.5}$$

---

*In the case of a gas, instead of c we write c_p, the specific heat at constant pressure.

where α is the heat transfer coefficient (units W/m^2K). The value of α depends upon the characteristics of the fluid, the flow conditions, and the dimensions of the solid body and its surroundings. Table 3.2.1 gives an indication of the order of magnitude of α and shows the enormous range of rates of heat transfer that can occur.

Table 3.2.1   Heat Transfer Coefficient α: Orders of Magnitude

	α, W/m^2K
Free and forced convection: Gas	10–100
Free and forced convection: Liquid	500–5000
Boiling water	up to 20 000
Condensing steam	up to 60 000

The concept of the temperature boundary layer is fundamental to an understanding of convective heat transfer. In this layer the temperature varies as a continuous function from that of the solid wall to that of the bulk of the fluid outside the boundary layer, Fig. 3.2.1. All bodies immersed in a fluid and subjected to either heating or cooling are surrounded by a thermal boundary layer, which has properties analogous to those of the boundary layer encountered in fluid flow. The calculation of temperature and velocity distributions in a thermal boundary layer on a purely theoretical basis is possible only in simple cases. It is generally necessary to resort to experiment.

It is frequently possible to apply the principles of similarity [11, p. 13] to the interpretation of these experimental results. Similarity theory makes possible the application of measurements made, for example, on a body of one size and with a particular flow velocity to the prediction of performance with bodies of other sizes, other flow rates and other fluids.

3.2.1   Dimensionless Numbers Associated with Free and Forced Convection

The application of similarity theory to convective heat transfer followed the development of the theory in the field of fluid mechanics [11]. In the course of development of the theory of heat transfer a large number of dimensionless groups have been proposed, of which the most important are the Nusselt, Prandtl and Grashof Numbers.

The Nusselt Number represents the convective heat transfer coefficient in a liquid or gas in dimensionless form:

$$(Nu) = \frac{\alpha l}{\lambda}$$

where λ = thermal conductivity of the fluid in the absence of convective heat transfer.

The Nusselt Number may be interpreted as the ratio of the actual rate of heat transfer $\dot{q}$ to the rate of heat transfer $\dot{q}_1$ that would occur as a consequence of conduction only through a layer of the fluid of thickness l (the characteristic length) under the same conditions of temperature difference, Fig. 3.2.2. In the immediate neighbourhood of the wall (laminar boundary layer or laminar sub-layer of a

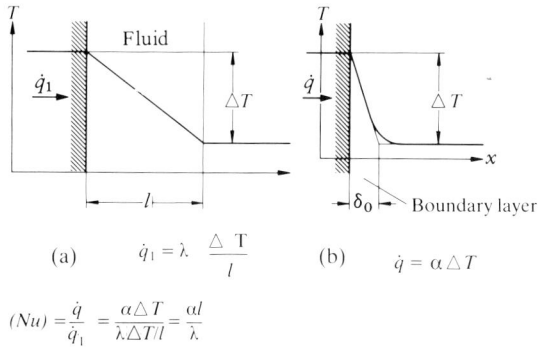

Fig. 3.2.2 Significance of the Nusselt Number:
(a) Heat transfer by pure conduction
(c) Heat transfer by conduction in boundary layer, then by convection

turbulent boundary layer) heat transfer does in fact take place as a consequence of conduction only. If we were to define the characteristic length by an appropriate boundary layer thickness, $l = \delta_0$ then we would write $\dot{q} = \dot{q}_1$ and $(Nu) = 1$. As, however, in most cases δ_0 is unknown, we choose l as some other characteristic length—for example, the diameter of a cylinder from which heat is being transferred—and determine the Nusselt Number by experiment. We may then write:

$$(Nu) = \frac{l}{\delta_0}$$

Temperature boundary layers are analogous to boundary layers in fluid flow and in gases they are of roughly the same thickness. The relation between the two thicknesses depends on the property of the fluid defined by the Prandtl Number:

$$(Pr) = v/a$$

where $v =$ kinematic viscosity and $a =$ thermal diffusivity.

To clarify the significance of the Prandtl Number, consider a laminar jet of fluid emerging at velocity v and temperature T into a body of the same fluid at rest but at a lower temperature T_0. As the jet penetrates further into the surrounding fluid its width increases, Fig. 3.2.3—a consequence of fluid friction and turbulent mixing—and the velocity and temperature profiles take up the form sketched in the figure. The Prandtl Number represents the ratio between the velocities of transverse propagation of

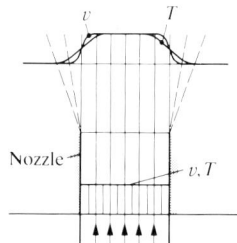

Fig. 3.2.3 Significance of the Prandtl Number

39

velocity and temperature. For air at atmospheric conditions $(Pr) = 0\cdot71$, implying that "temperature" diffuses into the surroundings more rapidly than "velocity". The velocity boundary layer is in consequence thinner than the temperature boundary layer; for $(Pr) = 1$ they would both have the same thickness.

In the solution of problems associated with free convection the buoyancy forces in a warmed fluid play an important part. The appropriate dimensionless group is the Grashof Number:

$$(Gr) = \frac{g\beta\Delta Tl^3}{v^2}$$

where g = gravitational acceleration, β = coefficient of cubical expansion of the gas, ΔT = characteristic temperature difference, l = characteristic length, and v = kinematic viscosity.

It may be shown that the Grashof Number is analogous to the Reynolds Number with the feature that the characteristic velocity is that developed by the buoyancy forces under laminar flow conditions.

3.2.2 The Temperature Boundary Layer

Heat leaving the surface of a solid body immersed in a fluid first encounters a layer of fluid at rest relative to the surface, Fig. 3.2.4. With increasing distance from the wall the velocity v parallel to the wall surface increases, and if the boundary layer is laminar there is no mixing of the fluid in successive layers. Heat is transmitted through the fluid by pure conduction, and the velocity and temperature distributions are as sketched in the figure.

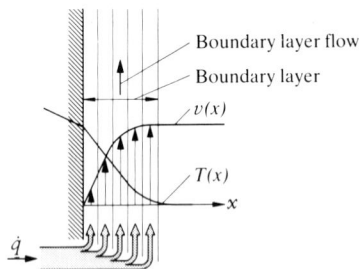

Fig. 3.2.4   The laminar boundary layer

If the boundary layer is turbulent the general direction of flow of the successive layers of fluid remains parallel to the wall surface, as in Fig. 3.2.5, but particles of fluid are caught up in the general turbulence of the flow and move irregularly in a transverse direction normal to the general direction of flow. (See [11], pp. 83–4 for a discussion of this phenomenon.) These fluid particles are a medium for the transfer of heat normal to the direction of flow of far greater intensity than that corresponding to pure conduction. Observation by Schlieren apparatus of the flow of air past a heated body arising from natural convection gives a clear indication of these motions.

40

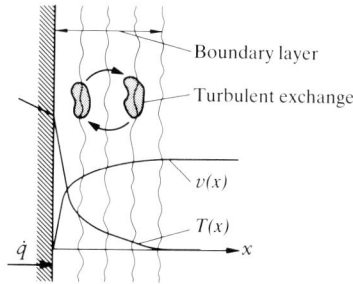

Fig. 3.2.5 The turbulent boundary layer

Fluid mechanics shows that, even when the boundary layer is generally turbulent, there is always a laminar sub-layer close to the body surface and in this layer heat transfer must take place solely due to conduction. This permits the calculation of the limiting temperature gradient at the surface by writing equation (3.1) in differential form:

$$\frac{dT}{dx} = \frac{\dot{q}}{\lambda}$$

where λ = thermal conductivity of the fluid. Substituting for $\dot{q}$ from equation (3.5) we may write:

$$\frac{dT}{dx} = \frac{\alpha}{\lambda} \varDelta T = \frac{\varDelta T}{k} \tag{3.6}$$

where $k = \lambda/\alpha$ has the dimension of a length and may be used in the graphical construction of the temperature profile, Fig. 3.2.6.

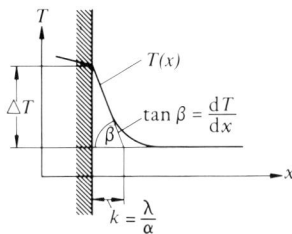

Fig. 3.2.6 Graphical construction of boundary layer profile

The simplest case of a thermal boundary layer is that arising from flow along a heated (or cooled) flat plate, Fig. 3.2.7. The boundary layer has zero thickness at the sharp leading edge of the plate, and its thickness then increases progressively with increasing distance from this leading edge. As already noted, the rate of heat transport normal to the surface depends primarily on the nature of the boundary layer, whether laminar or turbulent. It is usually observed that near the sharp leading edge of the plate the boundary layer is laminar, eventually becoming turbulent at a certain distance from the leading edge.

41

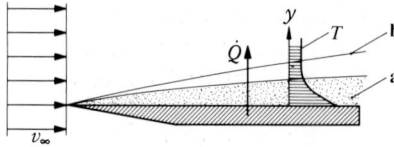

Fig. 3.2.7 Thermal boundary layer on a flat plate:
    a   Flow in the boundary layer
    b   Profile of boundary layer at a lower velocity v_∞

The analogy between heat transfer and fluid friction in this situation may be made clear, Fig. 3.2.8. Imagine two particles of fluid that change places due to the general turbulence in the boundary layer. In doing so they exchange momentum mv_1, mv_2 and also heat mc_pT_1, mc_pT_2. The momentum transfer tends to accelerate the flow of the fluid nearer the wall and reduce the velocity of that at a greater distance; it can be regarded as equivalent to a shear force τ. The heat transfer tends to increase the temperature of the fluid further from the wall and is equivalent to a heat flow $\dot q$ in a direction normal to the wall.

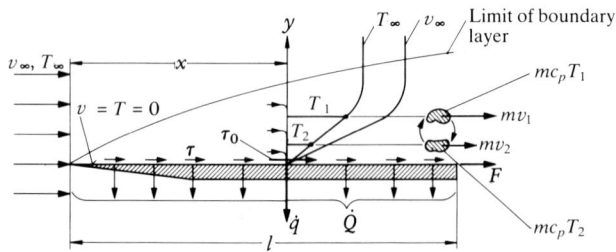

Fig. 3.2.8 Analogy between heat transfer and fluid friction

The transfer of momentum and heat takes place as a consequence of a multitude of small transfers of this kind. Table 3.2.2 describes the analogy in quantitative terms.

Table 3.2.2 Analogy between Friction and Heat Transfer

Friction		Heat transfer	
Free stream velocity	v_∞	Free stream temperature	T_∞
Local frictional shear stress	$\tau(x)$	Local heat transfer rate	$\dot q(x)$
Total friction force	F	Total heat transfer	$\dot Q$
Surface friction coefficient	C_f	Heat transfer coefficient	α
Transfer (loss) of momentum		Transfer of heat	
Laminar sub-layer $\tau = \mu \dfrac{dv}{dy}$		$\dot q = \lambda \dfrac{dT}{dy}$	

The full analogy between fluid friction and heat transfer must also take into account conditions in the laminar sub-layer, in which the shear force is a consequence of viscosity only while heat transfer takes place by conduction only. We find that here also the analogy holds, provided the dynamic viscosity μ and the thermal conductivity λ bear a specific relation one to the other. The required relationship is:

$$(Pr)=1 :(Pr)=\frac{v}{a}=\frac{\mu c_p}{\lambda}$$

since $\quad v=\dfrac{\mu}{\rho}\quad$ and $\quad a=\dfrac{\lambda}{\rho c_p}$

The analogy thus applies exactly for $(Pr)=1$ both for the laminar sub-layer and for the turbulent boundary layer. Many gases have Prandtl numbers in the region of 1 and the analogy applies reasonably well even with some departure from this value. This has the following important consequences:

(a) Resistance to fluid flow and heat transmission rate are proportional. An increase in the heat transfer rate by means of an increase in fluid velocity must be paid for by an increased pressure loss.

(b) The large quantity of research data available regarding flow resistance may be used as a basis for the calculation of heat transfer rates, for which in general less data is available.

The analogy is significant for flat plates, slender bodies, pipes and channels. In the case of bluff bodies the flow resistance is due less to laminar or turbulent shear forces than to separation and wake formation. The resulting forces, which can be large relative to those associated with boundary layer friction and turbulence, are not reflected in a corresponding increase in the rate of heat transfer.

The basic equation governing the analogy between flow resistance and heat transfer, for both laminar and turbulent flows with $(Pr)=1$ is

$$\frac{\alpha}{\tau_0}=\frac{c_p}{v_\infty}$$

where $\tau_0=$ local wall shear stress, $v_\infty=$ free stream velocity and c_p the specific heat of the fluid at constant pressure.

This relation is called the Reynolds Analogy. The analogy has been further developed to take into account values of Prandtl Number other than unity (Prandtl–Taylor modification of the Reynolds Analogy), [1]. To make use of the analogy to predict heat transfer coefficients we take values of τ_0 from the results of research in the field of mechanics of fluids. Table 3.2.3 summarizes the relationships derived in this way for flow along a flat plate:

Table 3.2.3 Heat Transfer, Flow along a Flat Plate

	α_x	$(Nu)_x$	$(Nu)_l$
Laminar boundary layer	$0\cdot332\,\lambda\sqrt{\dfrac{v_\infty}{x\cdot v}}\cdot(Pr)^{\frac{1}{3}}$	$0\cdot332\sqrt{(Re)_x}\,(Pr)^{\frac{1}{3}}$	$0\cdot664\sqrt{(Re)_l}\,(Pr)^{\frac{1}{3}}$
Turbulent boundary layer	—	—	$0\cdot057(Re)_l^{0\cdot78}(Pr)^{0\cdot78}$

Here x refers to the location distant x from the leading edge, l is the length of the plate, $(Nu)_x$ is calculated using the value α_x, while $(Nu)_l$ uses an average value of α for the whole plate.

In most practical applications we require the average heat transfer coefficient $(Nu)_l$ rather than the local value $(Nu)_x$. This is obtained by integration and is shown in the last column of the table. At the leading edge of the plate the heat transfer coefficient has a theoretical value of infinity; its variation with distance from the leading edge is sketched in Fig. 3.2.9.

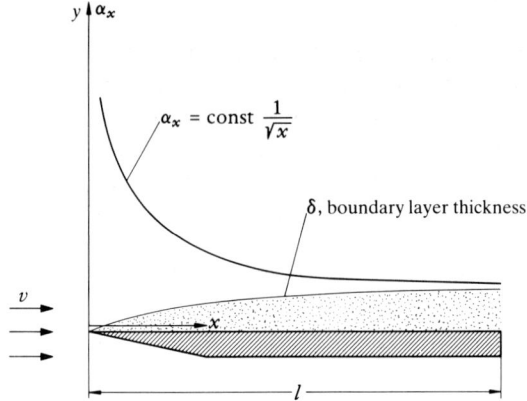

Fig. 3.2.9 Variation of heat transfer coefficient a with distance from leading edge

In the case of the laminar temperature boundary layer a complete theoretical solution, defining the temperature distribution at all points, is possible [1]. The transverse temperature profile at all stations downstream of the leading edge may be represented by a single dimensionless curve, Fig. 3.2.10. (The velocity distribution in a laminar boundary layer may be similarly presented) ([11], p. 50.))

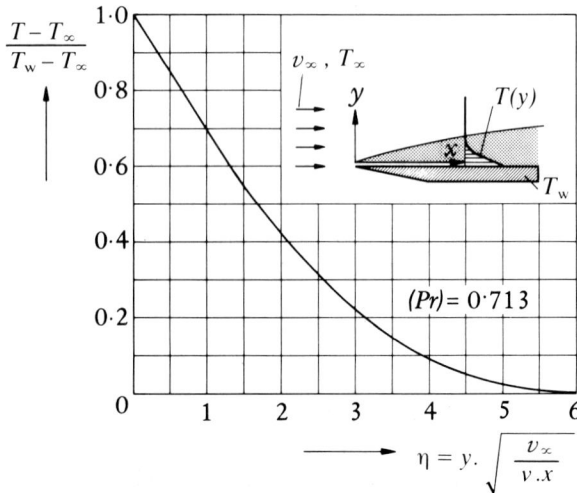

Fig. 3.2.10 Temperature distribution in laminar boundary layer for air, $(Pr)=0.713$: dimensionless presentation

There is a particular reason why the predictions of theory tend to be less exact in the field of convective heat transfer than in the field of fluid flow: in the case of flow phenomena in the absence of heat transfer, the characteristics of the fluid may in general be regarded as approximately constant; where temperature differences are involved this is not the case, since most characteristics of fluids are dependent on temperature. This feature also renders the application of results obtained with one fluid less accurate when applied to another fluid than is the case where constant temperatures only are involved.

Since the processes taking place in the boundary layer have a determining influence on convective heat transfer it is usual to insert in formulae values for such fluid properties as viscosity and density corresponding to the mean temperature between that of the solid boundary and that of the free stream remote from the boundary. The same rule must be used in applying the results of experiment to other situations.

At first sight it is surprising that the Reynolds Analogy holds for turbulent as well as for laminar flow since the characteristics of each are so different. The kinetic theory provides illumination at this point. According to the theory, fluid viscosity is a consequence of the diffusion of molecules in a direction perpendicular to the main flow. Molecules diffusing from a more rapidly flowing layer to one travelling at lower speed exchange momentum with those diffusing in the opposite direction. This implies the existence of a shear force τ_0:

$$\tau = \mu \frac{dv}{dy}$$

If a temperature difference exists perpendicular to the direction of flow, the diffusing molecules also carry heat in the form of internal energy with them. A molecule subject to diffusion thus carries with it momentum (= ordered motion in the direction of flow) and heat (= disordered molecular motion). The above description applies to laminar flow; in turbulent conditions the mechanism is similar except that, instead of individual molecules, we are concerned with elements of fluid of much greater size, each made up of a large number of molecules.

3.2.3 Flow Perpendicular to a Cylinder

As already indicated, observations of the flow pattern and pressure loss associated with flow past a cylinder cannot be applied by analogy to the corresponding heat transfer problem since so much of the flow resistance is associated with the negative pressure in the separated wake downstream of the cylinder. This resistance does not arise as a consequence of shear forces in the fluid. A boundary layer similar to that in the case of the flat plate exists only over that part of the cylinder forward of the separation point. There is no point corresponding to the leading edge of the flat plate where the boundary layer has zero thickness with the implication of a theoretically infinite rate of heat transfer. Even at the stagnation point the boundary layer has a finite thickness. This may be demonstrated in a very simple manner by making a cylinder of paper and attempting to ignite it with a match, when it will be found that the paper bursts into flame instantly when held vertically but only after an interval when held horizontally, Fig. 3.2.11.

Direct experiment, interpreted in terms of similarity theory, necessarily plays a

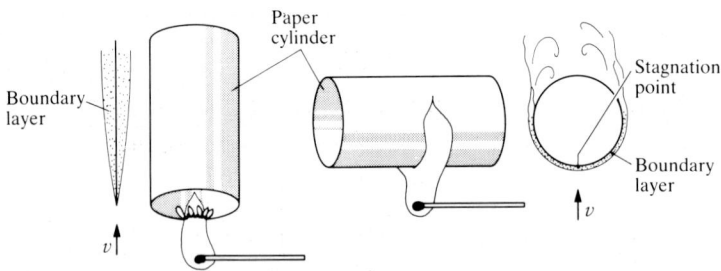

Fig. 3.2.11 Simple demonstration of thermal boundary layer thickness

large part in the acquisition of data concerning flow past bodies such as tubes. Fig. 3.2.12 shows the flow pattern past a cylinder and the corresponding local values of heat transfer coefficient, expressed as Nusselt Number and plotted on polar co-ordinates. For practical purposes only the average value of heat transfer coefficient is of significance. In general, for forced convection, it may be expressed in the form:

$$(Nu) = f((Re).(Pr))$$

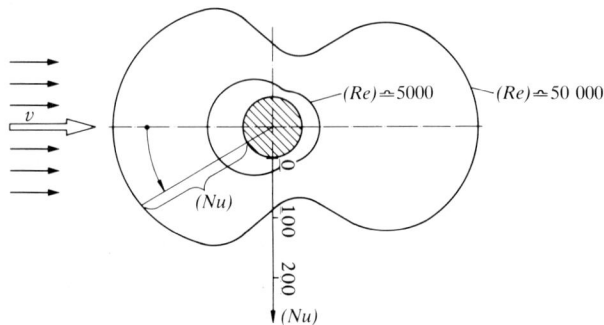

Fig. 3.2.12 Variation in surface heat transfer coefficient for a cylinder in cross flow

Fig. 3.2.13 [4] summarizes the results of a large number of measurements taken from the literature for the flow of air past circular tubes, cylinders and wires. There is an appreciable scatter in the results, associated with variations in such factors as the level of turbulence in the flow, roughness of the tube surface, and the direction of heat flow. The results shown in Fig. 3.2.14 may be summarized by an equation in the form:

$$(Nu) = k_1(Re)^{k_2} \qquad (3.7)$$

where k_1, k_2 are experimental constants.

Experiments with other fluids having other physical properties and different Prandtl Numbers lead to the generalized equation:

$$(Nu) = k_1(Re)^{k_2}(Pr)^{0.31} \qquad (3.8)$$

Around the forward or upstream tube surface which is covered with a boundary layer heat transfer takes place in an essentially similar manner to that described for

46

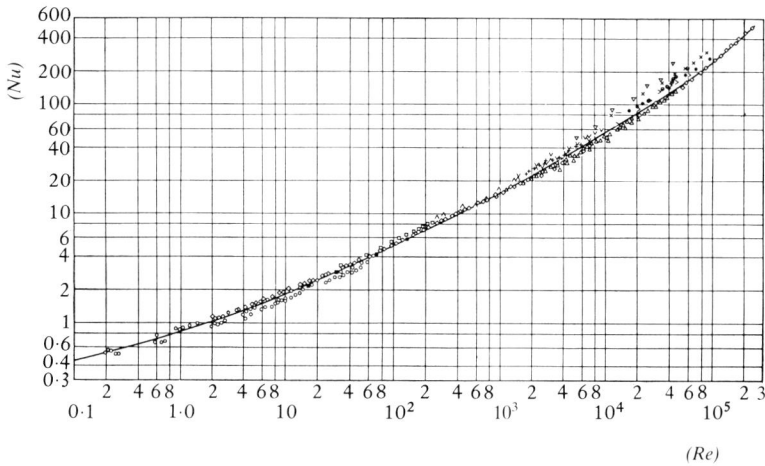

Fig. 3.2.13 Relation between Nusselt and Reynolds Numbers for isolated cylinder in cross flow

the flat plate. After the separation point the fluid in the boundary layer is rapidly mixed with that in the turbulent wake, with consequent equalization of temperature. Heat transfer also takes place downstream of the separation points between the tube surface and the fluid in the wake, again by way of a local, if disordered, boundary layer.

This applies to a single tube in isolation. In the case of a tube tank the rate of heat transfer tends to be higher and to rise as the flow advances through the bank, as a consequence of an increased level of turbulence.

Table 3.2.4 summarizes the equations governing heat transfer in the technically important case of flow of a fluid through a tube. A distinction must be made between

Table 3.2.4 Heat Transfer, Flow through Cylindrical Tube

Flow	Nusselt Number
Laminar	$$(Nu) = 3 \cdot 65 + \frac{0 \cdot 0668 \frac{d}{L}(Re) \cdot (Pr)}{1 + 0 \cdot 04 \left[(Re) \cdot (Pr) \cdot \frac{d}{L}\right]^{\frac{2}{3}}}$$
Turbulent	$(Nu) = 0 \cdot 032 (Re)^{0 \cdot 8} \cdot (Pr)^n \cdot \left(\frac{d}{L}\right)^{0 \cdot 054}$ for $(Re) > 10\,000$ $\quad$ $n = 0 \cdot 37$, heating $n = 0 \cdot 30$, cooling of fluid

laminar and turbulent flow, depending on the value of the Reynolds Number $(Re) < 2300$, laminar, $(Re) > 2300$, turbulent. The equation for laminar flow is particularly simple:

$$(Nu) = 3 \cdot 65 \qquad (3.9)$$

47

This value applies, however, only for "fully developed" flow. When a fluid enters a tube it must travel an appreciable distance, the transition or entry length, before the velocity distribution assumes a form that then persists unchanged throughout the length of the pipe, [11], pp. 55–6. The transition length for laminar flow is in the region of 60d, where $d=$ pipe diameter and between 10d and 40d for turbulent flow. The corresponding thermal transition length, beyond which the heat transfer rate departs by less than 1 percent from the value $(Nu) = 3.65$ is given by:

$$L_T = 0.05 . (Re) . (Pr) . d \tag{3.10}$$

In the case of oils, which may have Prandtl Numbers in the region of 100, the thermal transition length may be very large and cannot be ignored. In the case of turbulent flow the hydrodynamic and thermal transition lengths are of approximately the same length. The equations in Table 3.2.4 give mean values of Nusselt Number for the full length of the tube.

The velocity profile in a circular tube with fully developed laminar flow is parabolic. The corresponding temperature profile is similar, but with a somewhat sharper curvature at the centre of the tube. In turbulent flow the velocity profile is much flatter; for $(Pr) = 1$ the velocity and temperature profiles are identical, while for air, $(Pr) = 0.71$, they differ only slightly.

3.2.4   Free Convection from Horizontal Tube

An exact solution exists for the heat transfer by free or natural convection from a vertical plate of uniform surface temperature. The form of this equation influences those of the empirical equations which have been developed for bodies of other forms. In general, for free convection:

$$(Nu) = f [(Gr),(Pr)] \tag{3.11}$$

Theoretical consideration and experiments lead to the modification:

$$(Nu) = f [(Gr) \times (Pr)] \tag{3.12}$$

Fig. 3.2.14 shows experimental results from various sources. In the region $(Gr).(Pr) = 10^4$ to 10^9 the flow is laminar and above this value turbulent. The corresponding equations are as follows:

$$10^3 < (Gr).(Pr) < 10^9 \ : (Nu) = 0.56[(Gr).(Pr)]^{1/4}$$
$$10^7 < (Gr).(Pr) < 10^{12} : (Nu) = 0.13[(Gr).(Pr)]^{1/3}$$
$$10^9 < (Gr).(Pr) < 10^{12} : (Nu) = 0.021[(Gr).(Pr)]^{2/5}$$

For $(Gr)(Pr) < 10^4$ special considerations apply, see Section 3.4.

3.2.5   Boiling and Condensing Heat Transfer

So far we have considered heat transfer to and from fluids, either in the liquid or in the gaseous phase. When the heat transfer is associated with a change of phase the phenomena are much more complex and have not at present been fully explored. Fig. 3.2.15 represents schematically the progressive supply of heat to water. During the phase change bc from water to steam the temperature remains constant although the

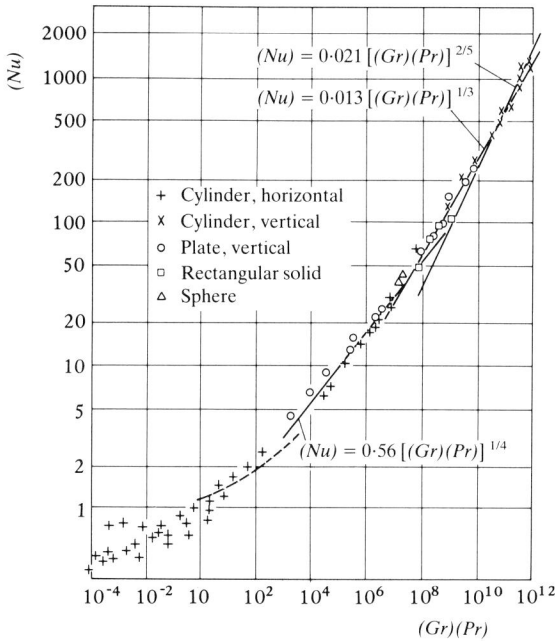

Fig. 3.2.14 Experimental results for Nusselt Number in free convection

quantity of heat supplied is relatively large. Exactly the reverse process takes place during condensation of steam to water.

Similarity theory has little to contribute to an understanding of the heat transfer processes taking place during boiling and condensation, neither is it possible to draw analogies with fluid flow processes. Generalized equations are not available and extensive experimental work is necessary.

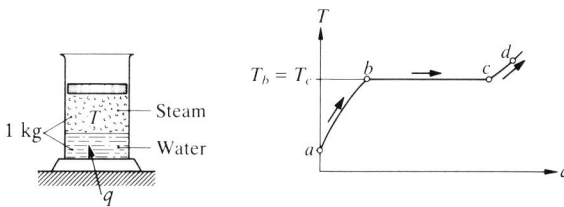

Fig. 3.2.15 The boiling process:
a–b Heating of liquid   b Start of boiling
c All liquid vaporized   c–d Superheating of vapour
                            $T_b = T_c$ boiling temperature

3.2.6 Boiling Heat Transfer [6]

The principal phenomena are best described with reference to a simple experiment involving boiling water and an electrical immersion heater. Knowing the electrical power supplied to the heater and its surface area, we may calculate the heat flux

49

$\dot{q}$ W/m². The results of such an experiment may be plotted in terms of $\dot{q}$ against the difference between the element surface temperature T_S and the boiling temperature of the water T_B, Fig. 3.2.16. It is convenient to adopt a logarithmic scale. We may distinguish four separate regimes, identified in the figure.

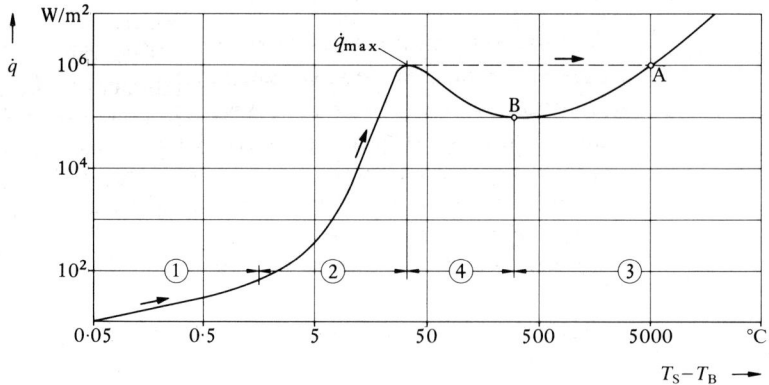

Fig. 3.2.16   Relation between rate of heat transfer in boiling and temperature difference $T_S - T_B$

1   Surface or Pool Boiling

The water vaporizes at the surface of the pool. The temperature of the heated element exceeds that of the bulk liquid by only a few degrees, giving rise to convection currents in the water and the occasional formation of vapour bubbles which break loose from the surface of the heating element and rise to the surface of the pool, Fig. 3.2.17. Such bubbles can only form when the temperature of the heating surface appreciably exceeds that of the bulk liquid because of the influence of surface tension. As a consequence the vapour pressure necessary to form a bubble must appreciably exceed the ambient pressure, with a consequent increase in the boiling temperature. The additional pressure arising from surface tension varies inversely with the bubble diameter.

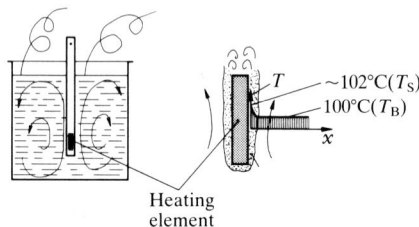

Fig. 3.2.17   Surface or pool boiling

The laws of natural convection apply in this regime and the rate of heat transfer is a function of the Grashof and Prandtl Numbers, equation (3.12). It is, however, of little technical interest since with a slight increase in the temperature of the heating element the rate of heat transfer becomes enormously greater.

50

2   Nucleate Boiling

With further increase in the element surface temperature vapour bubbles form directly on the element and rise to the surface of the liquid. Direct evaporation from the liquid surface plays a smaller part in the process. The rising bubbles generate an intense convective flow which sweeps the newly formed bubbles away from the heating surface. A tenfold increase in the difference in temperature between the element and the bulk liquid gives rise to an increase in the region of 10^3 to 10^4 times in the rate of heat transfer.

3   Film Boiling

With increasing $\dot{q}$ the intensity of nucleate boiling increases until eventually the water is no longer able to reach the surface of the heater between the bubbles. The surface of the heater then becomes covered with a continuous film of steam from which large bubbles become detached and rise to the surface of the water. Under film boiling conditions the heat transfer rate is significantly lower than when the liquid is able to reach the surface of the element. The point at which nucleate boiling reaches its greatest intensity, and prior to the onset of film boiling, corresponds to the maximum heat flux, $\dot{q}_{max}$.

In order to achieve the same peak rate of heat transfer $\dot{q}_{max}$ the temperature of the element must rise to that corresponding to point A, Fig. 3.2.16. With water, the temperature at A is so high that the element will melt, but with some organic fluids boiling at a much lower temperature than water it is possible to extend the curve of Fig. 3.2.16 beyond point A. The process of transition from maximum rate of nucleate boiling to film boiling is known as "burn-out". If the heating element has not burned out it is possible, after reaching point A, to reduce the rate of heat input and achieve stable operation in the zone between point A and point B.

4   Unstable Regime

Under this regime a mixture of nucleate and film boiling takes place, with rapid and irregular changes in the location of the film and nucleate boiling areas. It is not possible to explore this region with an apparatus of the type shown in Fig. 3.2.17. An alternative arrangement, in which surface temperature rather than rate of heat flux is the controlled variable, is used in Experiment 3 below and permits the unstable regime to be investigated.

Film boiling is of little technical interest, but the process of nucleate boiling and in particular the precise value of $\dot{q}_{max}$ are of great technical importance. It is particularly vital in the field of atomic power that burn-out should not occur in water-cooled reactors, otherwise melting of the fuel elements will follow. The value of $\dot{q}$ is a function of pressure and, in the case of boiling taking place within tubes, of the velocity of flow and the proportion of the volume occupied by steam bubbles.

Point B, Fig. 3.2.16 corresponding to the minimum rate of heat flux for stable film boiling is known as the Leidenfrost point. The so-called Leidenfrost Phenomenon is observed when a drop of water is spilled on to a horizontal surface, the temperature of which equals or exceeds that of point B. The water droplet is then supported on a cushion of steam, by which it remains suspended out of contact with the surface until it is completely evaporated.

The transition from pool boiling to nucleate boiling has become significant in connection with pressurized water reactors; in this boiling regime, bubbles may form on the hot surface, detach themselves and subsequently condense in the surrounding liquid, giving a high rate of heat transfer without the onset of actual boiling and with modest temperature differences.

In conclusion, it is worth commenting once more on the enormous rates of heat transfer achieved by the boiling process. Fig. 3.2.16 shows a value of $\dot{q}_{max}$ of 10^6 W/m^2. With increasing pressure $\dot{q}_{max}$ increases further and at a pressure of about 80 atmospheres reaches a maximum three or four times as great as that at atmospheric pressure. Such rates of heat transfer correspond to an energy input of about 1 kW per square inch of element surface.

3.2.7   Condensing Heat Transfer

If vapour is brought into contact with a surface the temperature of which is less than that corresponding to the boiling point of the vapour at the prevailing pressure, condensation takes place with transfer of the corresponding latent heat of evaporation to the wall. It is usual to arrange for condensation to take place on vertical walls or on arrays of horizontal tubes, so that the condensate runs down the surface and hinders the process of heat transfer to the minimum extent. The simplest case concerns a vertical plate, Fig. 3.2.18. The condensed liquid forms a layer of increasing thickness as it runs down the surface and the flow conditions in the film resemble those in a boundary layer. Nusselt developed a successful theoretical treatment of the phenomenon of film condensation.

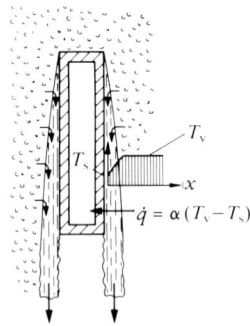

Fig. 3.2.18   Film-wise condensation on a vertical surface

On occasion, experiments indicate a significantly greater rate of heat transfer than that predicted by the Nusselt theory; under particular circumstances, the vapour does not condense in the form of a continuous film but as individual drops, which grow and individually run down the wall surface, Fig. 3.2.19. Clean steam condensing on a clean surface always forms a film, but the presence of certain oils promotes drop-wise condensation. As a basis for design it is safer to assume film-wise condensation. Table 3.2.5 gives an indication of the rates of heat transfer to be expected for these modes of condensation. For film-wise condensation the rate is only about one-eighth of the maximum for the reverse process of boiling at atmospheric pressure and one-thirtieth of the maximum at elevated pressures.

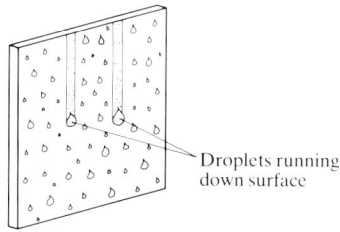

Fig. 3.2.19 Drop-wise condensation on a vertical surface

Table 3.2.5 Condensing Heat Transfer from Water

	α, W/m^{2}K	$\varDelta$T, K	$\dot{q}$, W/m^2
Drop-wise condensation	~60 000	10	~600 000
Film-wise condensation	~12 000	10	~120 000

The presence of even small quantities of gas in the condensing vapour results in a very large reduction in the rate of heat transfer. This is why it is necessary, in steam power plants, to de-aerate the feed water and to fit the condenser with an air pump.

3.3 Radiation [7]

Solids, liquids and to some extent gases are capable of transmitting heat not only as a consequence of direct contact but in the form of radiation. Infra-red radiation, which occurs down to the lowest temperatures, differs from light only because of its longer wavelength. The participation of thermal radiation in a thermodynamic process has no effect upon the validity of the First and Second Laws of Thermodynamics. In particular the effects of thermal radiation do not make possible the construction of a heat engine with efficiency greater than that of the Carnot cycle. It is self-evident that a hot body radiates with a greater intensity than a cooler one, with the consequence that radiation, like other modes of heat transfer, tends towards the equalization of temperatures. Heat transfer by radiation may be of practical significance at quite low temperatures, far below the temperature, roughly 600°C, at which a body first emits visible radiation. An ordinary water-filled "radiator" transfers a substantial proportion of the heat it transmits to its surroundings by radiation, see Experiment 4 (p. 79).

When radiation impinges upon a body, part is reflected and part is absorbed, giving rise to an increase in internal energy. In the case of transparent bodies part of the radiation may also be transmitted. The laws of radiation may be particularly simply stated for a body the surface of which has the characteristic of reflecting no energy, absorbing the whole of that falling upon it. Such a body is known as a black body, although a body capable of absorbing nearly all the thermal radiation falling upon it may yet not absorb all incoming light and hence not appear black to the eye. In reality no surface can be regarded as completely black, although some, for example a metal surface covered with lamp-black, approach the ideal closely.

53

Although the laws governing radiation from real bodies are complex, that defining the radiation from an ideal black body is relatively simple. The Stefan–Boltzmann Law states that:

$$\dot{Q} = A\varepsilon\sigma \left(\frac{T}{100}\right)^4 \tag{3.13}$$

where A = surface area, ε = emissivity relative to that of a black body, for which $\varepsilon = 1$ by definition, σ = Stefan–Boltzmann constant = 5·77 W/m^2K^4 and T = absolute temperature.

Usually a body also receives radiant energy from the surroundings. For a body at temperature T_1 enclosed in a space substantially greater in volume than the body, and having a containing surface of temperature T_2, Fig. 3.3.1, the following relationship applies:

$$Q = A_1\varepsilon\sigma \left[\left(\frac{T_1}{100}\right)^4 - \left(\frac{T_2}{100}\right)^4\right] \tag{3.14}$$

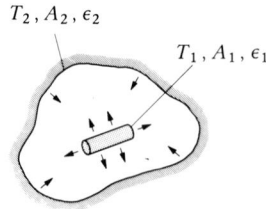

Fig. **3.3.1** Exchange of radiation between a body and its surroundings

In the case of a hot body located in a gas-filled space, for example a hot-water radiator in a room, heat transfer takes place both by radiation and by free convection. As we have seen, the laws governing these two methods of heat transfer are very different in form. In order to investigate radiation it usually necessary to evacuate the space or at any rate to reduce the gas pressure as far as possible.

3.3.1   Solar Radiation

The "solar constant", or the rate at which energy is received from the sun at a distance corresponding to the mean radius of the earth's orbit, is approximately 1350 watts/metre². This rate is not a true constant but rarely varies more than ±4 percent from the mean value, an insignificant variation compared with that due to the earth's atmosphere. Ultraviolet radiation is almost completely absorbed by the atmosphere, and about half the radiation energy received at sea level is within the visible range of wavelength, the remainder being within the infra-red wavelength region.

The proportion of the solar radiation absorbed by a clear sky is a function of the thickness of air through which the radiation has to pass; evidently this is at a minimum with the sun overhead and increases as the sun approaches the horizon. An analysis of the relationship between the direct radiation absorbed and the elevation (zenith distance) of the sun is given in [16].

54

Under clear sky conditions the level of incident energy at the earth's surface may range from about 0·6 to 0·8 × solar constant corresponding to a range from 800 W/m² to 1100 W/m². "Standard sunlight" is sometimes taken as 1 kW/m², but the highest level of radiation which is likely to be encountered in the latitude of the United Kingdom is about 950 W/m², while a good average figure for bright sunshine in temperature latitudes would be 700 W/m².

Fig. 3.3.2 shows the theoretical level of solar radiation on a horizontal plane for a clear sky at various latitudes in the Northern Hemisphere throughout the year. It will be seen that at midsummer it is possible to receive up to 8 kWh/m² per day almost independent of latitude, but the amount received in midwinter falls dramatically with increasing latitude. This curve illustrates one of the most serious practical limitations to the use of solar energy: in all but tropical latitudes little energy is available at the time of the year when it is most needed.

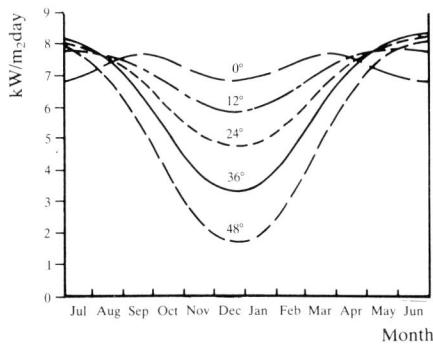

Fig. 3.3.2   Clear day solar radiation on a horizontal plane at various latitudes

Fig. 3.3.3 shows the spectrum of solar energy at the mean distance of the earth from the sun in the absence of intervening atmosphere. This curve corresponds very closely to the radiation from a black body at a temperature of 6000K. The wavelength range of the visible spectrum, from 0·38 to 0·78 μm, is shown.

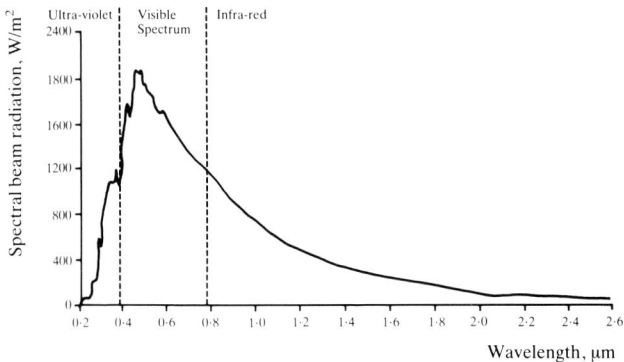

Fig. 3.3.3   Spectrum of solar energy at the mean distance of the earth

55

3.3.2 Solar Energy Collectors

The search for alternative sources of energy will be a major preoccupation of the next generation of engineers and scientists. Solar energy can be expected to make a modest contribution, perhaps 2 percent of the energy needs of the United Kingdom by the year 2000, according to one prediction [17].

Solar energy collectors may be divided into two broad categories: flat-plate collectors and focusing collectors. It is probable that the great majority of applications of solar energy in the foreseeable future will make use of flat-plate collectors, which have the advantages of simplicity and of the absence of the necessity for a steering or solar tracking mechanism.

Such collectors are already widely used for supplying hot water for domestic use and for the heating of swimming pools. A typical heater comprises a flat plate to which a series of tubes are metallically bonded. The surface of the plate that faces the sun is coloured matt black and the plate is enclosed in an airtight box having a glass cover. Heat loss from the rear surface of the plate is minimized by a thick layer of insulating material.

The presence of the glass cover-plate is essential to the working of the collector. Although the glass reflects and absorbs some of the radiant energy, the reduction in heat loss from the collector plate exceeds the loss in transmission through the glass. The benefit arises mainly from the reduction in convective heat loss from the collector surface, but there is also a reduction in loss due to the so-called "greenhouse" effect. This arises because the glass is comparatively transparent to the short-wave radiation received by the collector, but opaque to the long-wave reflected radiation from the collector surface.

Fig. 3.3.4 Experimental flat-plate solar energy collector

Instruments used for the measurement of the intensity of solar radiation are known as solarimeters. The frosted glass dome solarimeter consists of a hemispherical glass casing enclosing a series of thermocouple junctions connected to a series of silver segments coloured alternately black and white. The apparatus is calibrated and gives an output proportional to the total incident radiation.

Fig. 3.3.4 shows an experimental flat-plate solar energy collector used by the authors, and Fig. 3.3.5 shows a performance curve; this is typical of the performance of collectors of this general type. The test was taken in the United Kingdom on a clear day in early summer and the average value of incident radiation, 650W/m², was typical of these conditions. The curve indicates very clearly the limitations of the device. The efficiency falls very rapidly with increasing temperature and, for the present apparatus, reaches zero when the mean circulating water temperature exceeds that of the surroundings by 57°C. The collector is ineffective as a source of heat at even moderately high temperatures and, indeed, is only 50 percent efficient at a temperature difference of 33°C. More elaborate designs, incorporating multiple glass cover-plates and specially treated collector surfaces, have been developed. However, the practicable working temperatures even for advanced designs at high levels of incident radiation are limited to about 90°C.

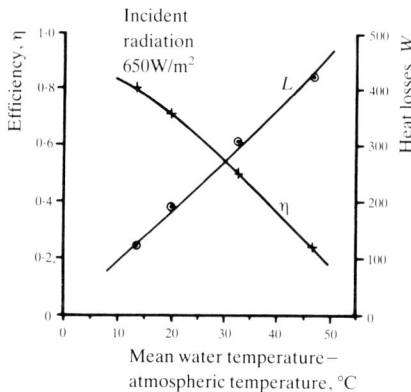

Fig. 3.3.5 Flat plate solar energy collector: relation between Efficiency, Losses and Temperature

3.4 Heat Exchange between Solid Bodies and Gases of Very Low Density [5]

It is frequently of practical importance to know how the heat transfer coefficient in free or forced convection changes with decreasing gas pressure. One might assume that α would be approximately proportional to pressure but in fact we find that, at least down to extremely low pressures, the latter has very little influence on the value of α. This may be clarified by reference to the kinetic theory. We have seen that heat transfer from a solid to a gas takes place as a consequence of the impact of gas molecules upon the surface and their reflection with increased kinetic energy. As the gas pressure is reduced, the number of molecules striking the surface falls, but the mean free path of the molecules increases inversely with the pressure. With reducing pressure the number of molecules drawing heat from the surface thus falls, but the velocity with which the transferred heat is diffused away from the surface increases. These two effects cancel out so that in theory the rate of convective heat transfer is independent of pressure. In practice the reduction with falling pressure is slight.

When the pressure becomes very low the situation changes. When the mean free path of the gas molecules becomes comparable to the thickness of the boundary layer or even to some characteristic length l of the solid body, it is no longer possible to speak of a convective gas flow in the usual sense. The phenomenon is characterized by a new dimensionless group, the Knudsen Number:

$$(Kn) = \frac{\lambda}{l}$$

where λ = mean free path and l = characteristic length.

The kinetic theory of gases permits the calculation of λ. For air,

$$\lambda = 159 \times 10^{-9} \frac{T}{h_0} \text{ metre,}$$

where T = absolute temperature and h_0 = absolute pressure, mmHg.

Fig. 3.2.15 showed the relationship between Nusselt Number and the product of Grashof and Prandtl Number in free convection. Results are given for $(Gr).(Pr)$ in the range 10^{-4} to 10^3, but no equation is shown. These results refer either to bodies having a very small characteristic length, for example fine wires, or situations in which the gas pressure is extremely low. The following expression for heat transfer from horizontal cylinders by free convection at low gas pressures is found in the literature [4]. This equation is empirical, but its derivation is supported by theory:

$$\frac{2}{(Nu)} = \log_e \left\{ 1 + \frac{6\cdot82}{[(Gr).(Pr)]^{1/3}} \right\} + (Kn) \left[\frac{8\gamma}{a(\gamma+1)} \right] - \log_e \left[1 + 2(Kn) \right] \quad (3.15)$$

where $\gamma = c_p/c_v$ and a is an "accommodation coefficient". This is the ratio of the actual energy exchange that takes place between a gas molecule and a surface at higher temperature upon which it impinges, to the maximum possible energy exchange. It may be taken as $0\cdot96$ for a matt black surface. This equation has validity in the range $(Gr).(Pr) = 10^{-8}$ to 10^{-1}.

3.5 Experiment 1: Thermal Boundary Layer on a Flat Plate

A study of the temperature boundary layer is fundamental to an understanding of convective heat transfer. The purpose of the present experiment is to study the simplest case: the temperature boundary layer on a flat plate along which a gas is flowing. The authors made use of a simple wind tunnel, Fig. 3.5.1. It is of the open-circuit type in which air is drawn into the working section by way of a shaped contraction in which acceleration of the flow takes place. The working section is followed by a diffuser in which the air velocity falls with some recovery of pressure, a fan, and a butterfly valve by which the velocity in the working section may be varied. An open-circuit wind tunnel is much cheaper and more compact than the closed-circuit type generally employed for research, but has the disadvantage that the flow

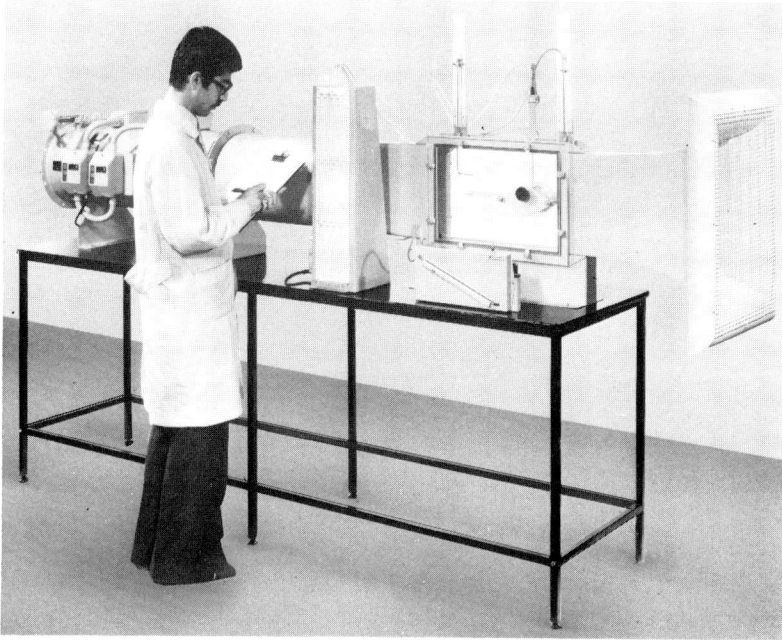

Fig. 3.5.1 Open-circuit wind tunnel

conditions in the working section may be disturbed by the presence of air currents and turbulence in the laboratory. The maximum velocity in the working section is 45 m/s (160 km/h).

A heated flat plate of aluminium alloy, having a sharpened leading edge, is mounted in the working section and may be traversed in a longitudinal direction by a rack and pinion mechanism. An electrical resistance heating element, the distribution of which is graduated to minimize longitudinal temperature gradients in the body of the plate, is bonded to the lower surface, Fig. 3.5.2.

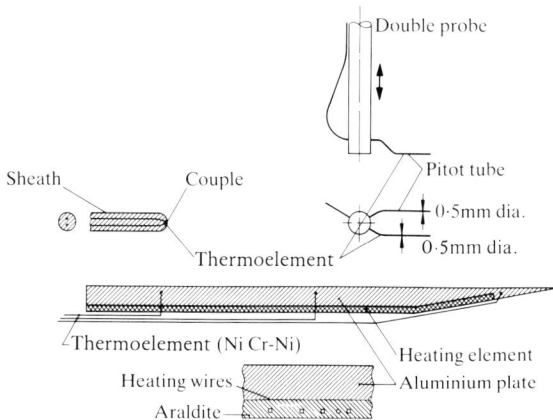

Fig. 3.5.2 Arrangement of thermal boundary layer experiment

Velocity and temperature profiles normal to the plate surface are explored by a double probe, comprising a pitot tube and a shielded chromel–alumel thermocouple mounted side by side and traversed by a common micrometer head. Both probes have a diameter of 0·5mm. The mean plate temperature is measured by three thermocouples embedded at points distributed along the plate length. Using the plate traversing mechanism and the micrometer head it is possible to take temperature and velocity profiles in transverse planes located at any desired distance from the leading edge of the plate.

A pen recorder having a full scale range of about 2mV is the most suitable means of observing the boundary layer temperature distribution. The air velocity is determined by measuring the stagnation pressure with the pitot tube, using an inclined water manometer connected between the pitot tube and a static tapping in the wind tunnel wall.

Since the diameter of the probe is 0·5mm the boundary layer thickness is given by:

$$\delta_T = y_1 - y_2 + 0\cdot25\,\text{mm},$$

where y_1 and y_2 are the micrometer readings at the extremity of the boundary layer and with the probe in contact with the plate, respectively.

The velocity of flow remote from the surface of the plate is measured by traversing the probe to a distance of at least 50mm from the plate surface and observing the pressure difference. v_∞ is then calculated in accordance with the following equation:

$$v_\infty = 75\cdot04 \sqrt{\frac{hT}{p_0}}$$

where h = stagnation pressure, mmH$_2$O, T = absolute temperature and p_0 = barometric pressure, N/m^2 (see [11], Appendix 1, for derivation).

Observations of the stagnation pressure are rather tedious as, owing to the small bore of the tube, the manometer readings take several minutes to stabilize.

The thermocouple probe is connected to the pen recorder and it is convenient to treat one of the thermocouples embedded in the plate as the cold junction, since we are only interested in the temperature difference between the probe and the plate, Fig. 3.5.3. The reading of the pen recorder corresponding to zero temperature difference is established by running the apparatus for a time with the heating element switched off. For the small temperature differences with which we are concerned it is acceptable to regard the relation between thermoelectric voltage and temperature difference as linear (for chromel–alumel, $\Delta T = 10°C$ corresponds to a change in e.m.f. $\Delta e_{10} \simeq 0\cdot4\,\text{mV}$).

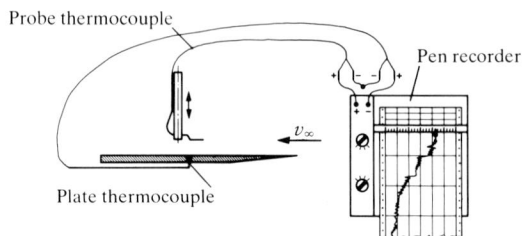

Probe thermocouple

Pen recorder

v_∞

Plate thermocouple

Fig. 3.5.3 Method of recording boundary layer temperature

3.5.1 Measurements and Calculations

A study of the thermal boundary layer falls into two parts: a determination of the profile of the boundary layer and an exploration of the temperature distribution in the layer at a number of transverse planes at various distances from the leading edge. These traverses cover the laminar, transition and turbulent regions of the corresponding velocity boundary layer.

The outer extremity of the thermal boundary layer is surprisingly clearly defined; at a sufficient distance from the plate surface the thermocouple probe reading is almost perfectly steady, but as the probe is traversed towards the plate a point is reached at which temperature fluctuations suddenly become apparent. This point corresponds to the limit of the thermal boundary layer, and is more easily estimated than the corresponding limit for the velocity boundary layer.

The variations in temperature due to turbulence in the boundary layer take place so rapidly that the 0·5 mm diameter thermocouple is not capable of following them (for measurements of this kind a hot-wire anemometer, in which a fine electrically heated wire is exposed to the turbulent flow, is capable of following the disturbances). The high-frequency variations are, however, accompanied by instabilities of appreciably longer period and these may be detected with the thermocouple probe, which has a time constant in the region of 0·5 second (see Fig. 2.8.6). A typical pen recorder is also able to follow oscillations of period of this order. Fig. 3.5.4 shows two typical recordings, one taken with a laminar boundary layer and one with a turbulent layer. After a little experience it is easy to decide whether the probe is lying within a laminar or a turbulent region.

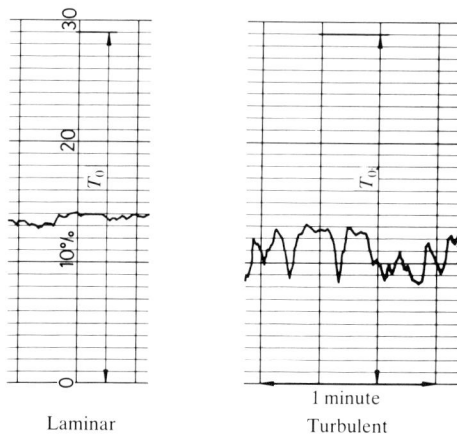

Fig. 3.5.4 Temperature variations in boundary layer

A complete analysis of the phenomenon involves both a study of the temperature boundary layer and, in parallel with it, a study of the velocity boundary layer. In what follows we concentrate on the former, referring the reader to [11], Experiment 4, for an account of a study of the velocity boundary layer carried out with essentially the same apparatus.

Temperature profiles are best taken starting with the probe in contact with the plate and moving in steps of 0·1 or 0·2 mm until the amplitude of the temperature variations indicated by the recorder falls, showing that the limit of the thermal boundary layer has been reached. It is necessary to observe the recording for an appreciable period for each position of the probe in order to make an accurate estimate of the mean temperature. Table 3.5.1 shows measurements of a typical temperature profile at a distance $x = 100$ mm from the leading edge with a velocity $v_\infty = 8·2$ m/s. This and other temperature profiles are shown in Fig. 3.5.5.

Table 3.5.1 Experiment 1: Thermal Boundary Layer on a Flat Plate: Observations

y	e.m.f.	$t_\mathrm{w} - t$	$\dfrac{t - t_\infty}{t_\mathrm{w} - t_\infty}$	$\eta = y \cdot \sqrt{\dfrac{v_\infty}{v \cdot x}}$
mm	mV			
0·25	0·09	2·3	0·86	0·58
0·45	0·20	5·0	0·71	1·04
0·65	0·31	7·8	0·54	1·50
0·85	0·40	10·0	0·41	1·97
1·05	0·45	11·3	0·34	2·43
1·25	0·53	13·3	0·22	2·89
1·45	0·55	13·8	0·19	3·35
1·65	0·59	14·8	0·13	3·81
1·85	0·62	15·5	0·09	4·28
2·05	0·64	16·0	0·06	4·74
2·25	0·65	16·3	0·04	5·20
2·45	0·66	16·5	0·03	5·66
2·65	0·67	16·8	0·01	6·13
2·85	0·68	17	0	6·60

$$x = 100 \text{ mm}$$
$$t_\infty = 22·5°C$$
$$\text{barometer} = 740 \text{ mmHg}$$
$$h = 4·00 \text{ mmH}_2\text{O}$$
$$v = 8·2 \text{ m/s}$$
$$v_\infty = 15·34 \times 10^{-6} \text{ kg/ms}$$
$$t_\mathrm{w} - t_\infty = 17°C$$
η, dimensionless thickness, see Fig. 3.2.10.

A complete study involves the collection of a considerable quantity of data, and for completeness a set of observations of the velocity boundary layer, in accordance with the procedure of [11], Experiment 4, should also be carried out.

The boundary layer profile is best observed with a fairly high stream velocity $v_\infty \sim 35$ to 45 m/s. Fig. 3.5.6 shows such a profile, together with examples of pen

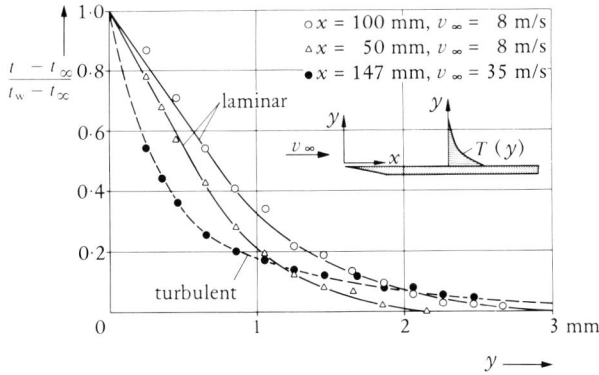

Fig. 3.5.5 Thermal boundary layer profiles, flat plate

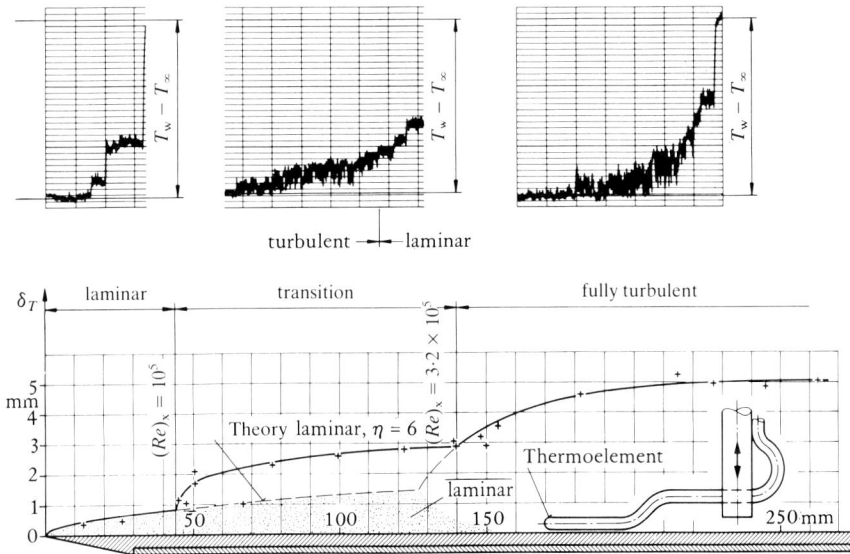

Fig. 3.5.6 Thermal boundary layer on a flat plate, $v_\infty = 35$ m/s

recordings illustrating the onset of temperature fluctuations as the boundary layer is entered. The corresponding Reynolds Numbers are also shown.

This is an instance in which it may be appropriate for several groups of students each to take a limited number of sets of results covering some part of the range of air velocities, heating rates, temperatures and velocities that require exploration. All the results may then subsequently be assembled in a single report.

3.5.2 Discussion of Results

Fig. 3.5.6 shows clearly the discontinuities associated with the transition from laminar to turbulent conditions. At the point corresponding to the onset of the transition region the Reynolds Number has the value $(Re)_x = 10^5$; at the later transition to fully turbulent conditions, also indicated by a discontinuity in the profile, $(Re)_x = 3 \cdot 2 \times 10^5$. These values agree well with those expected for transition in the corresponding velocity boundary layer ([11], Chapter 5).

Fig. 3.5.5 shows that the laminar temperature profiles are linear in the region of the wall, as the theory suggests, while the turbulent profiles show appreciable curvature up to the observation taken at the nearest available approach to the wall; special experimental techniques are required to explore this inner region. The laminar profiles are in fact calculated in accordance with the curve of Fig. 3.2.10, taking $\eta = 6$ as corresponding to the observed outer boundary of the layer. The fit of the experimental points to the curves is seen to be good.

If we extrapolate the straight line part of the laminar temperature profiles to intersect the axis we may determine the distance k. The analysis given earlier, Fig. 3.2.6, and equation (3.6), showed that the local heat transfer coefficient α_x is related to the value of k by the expression.

$$\alpha = \lambda/k$$

where λ = thermal conductivity of the gas (air in the present case), taken at the mean temperature of the plate surface and the free stream. Taking $\lambda = 0 \cdot 0264 \text{W/mK}$, we arrive at the values of local heat transfer coefficient α_x given in Table 3.5.2.

Table 3.5.2 Local Heat Transfer Coefficients, Flat Plate

x mm	v_∞ m/s	k mm	α_x W/m^2 K	$(Nu)_x$	$(Re)_x$	$(Nu)_{x\,theor}$
50	8·0	1·12	23·6	44·7	25000	46·8
100	8·2	1·43	18·5	70·0	51200	67·0

The local Nusselt Number is calculated very simply:

$$(Nu)_x = \frac{\alpha_x x}{\lambda} = \frac{(\lambda/k) x}{\lambda} = \frac{x}{k}$$

Both (Nu) and (Re) are calculated taking the distance x from the leading edge as the characteristic length. The values of (Nu) predicted from the theoretical equation of Table 3.2.3 are seen to be in good agreement with the values calculated from temperature profiles.

A similar extrapolation in the case of the turbulent profile depends upon the estimated form of the curve and is uncertain; the thickness of the laminar sub-layer is very much less than the minimum distance, 0·25 mm, from the wall surface at which the temperature profile may be observed.

It should be noted particularly that the whole of the heat transferred from the plate surface is contained in the boundary layer; none has found its way into the free stream since the temperature at the surface of the boundary layer remote from the plate is by definition equal to that of the free stream; there is thus no temperature difference to promote a heat flow. The heat will only be transferred from the boundary layer by large-scale mixing downstream of the plate.

64

3.5.3 Further Experiments and Questions

(a) A further exploration of the velocity distribution in the boundary layer in the laminar, turbulent and transition regions, comparison of results with accepted values, and application of Reynolds Analogy to a comparison of the velocity and temperature boundary layers.

(b) Why, despite their greater thickness, do turbulent boundary layers exhibit a higher heat transfer coefficient?

(c) By attaching a transverse trip wire just downstream of the leading edge the entire boundary layer may be made turbulent. Investigate the effect on boundary layer profile and rate of heat transfer.

3.6 Experiment 2: Heat Transfer by Forced Convection from a Single Tube and a Tube Bank

The experimental measurement of heat transfer coefficients is not a simple matter. Experiment 1, while it demonstrates many features of heat transfer from a flat plate, does not permit overall measurements of the heat transfer coefficient. To do so it would be necessary either to attempt to prevent heat loss from all the surfaces of the plate other than that under investigation or we should need to "calibrate" the plate by measuring the temperature gradient normal to the surface at a number of points, making use of the thermal conductivity of the material to calculate the corresponding rate of heat transfer from the surface. We are thus involved in a number of accurate temperature measurements; in addition it would be necessary to measure accurately the air flow rate past the plate and the initial and final temperatures of the air after full mixing between the boundary layer and the undisturbed air had taken place. In addition, substantial edge-effects, such as heat loss from the edges of the plate and the walls of the wind-tunnel working section, would have to be taken into account.

Similarly, it is extremely difficult to measure the heat transfer coefficient from the surface of a tube by passing a hot fluid through the tube, measuring its heat loss and also measuring the flow rate and rise in temperature in the stream passing the tube. End-effects are particularly important in experiments of this kind. When designing experiments intended to measure rates of heat transfer it is necessary to look particularly critically at the experimental set-up, the methods of measurement, and the possible sources of error.

An experimental technique for measuring the rate of heat transfer from the surface of a cylinder that avoids most of these difficulties is illustrated in Fig. 3.6.1 and 3.6.2. Air is drawn through a channel of rectangular cross-section into which cylindrical perspex rods may be inserted transversely to model a single tube, a row of tubes or a tube bank. At any desired position in this array of rods an element of the form shown in Fig. 3.6.3 may be inserted. This comprises a rod of high purity copper sandwiched between two plastic end-members. A copper–constantan thermocouple is embedded in the centre of the copper cylinder and the experimental technique is to heat the cylinder to about 70°C, insert it in the duct, and plot the cooling curve using a pen recorder.

Fig. 3.6.4 shows the principal dimensions of the working section of the apparatus

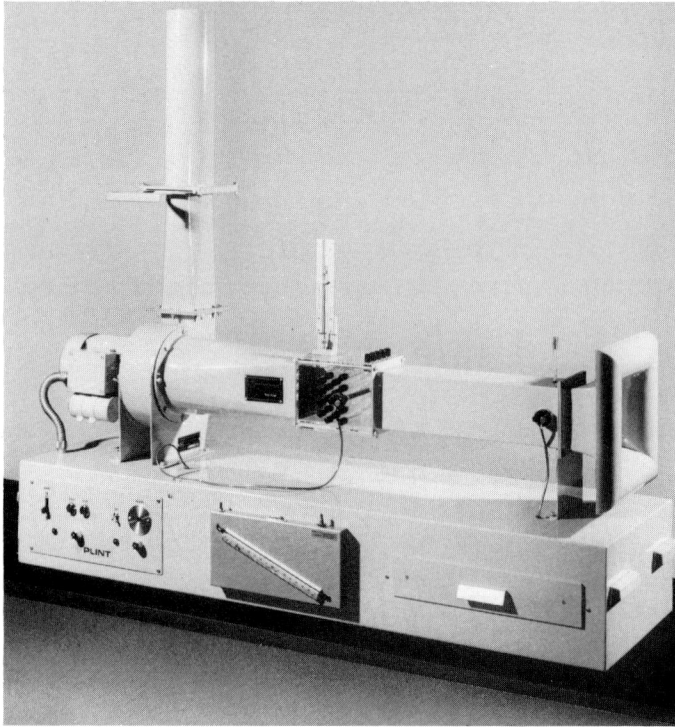

Fig. 3.6.1 Cross-flow heat exchanger

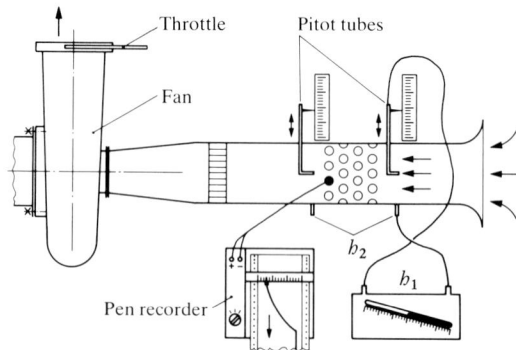

Fig. 3.6.2 Schematic arrangement of cross-flow heat exchanger

used by the authors. The tube bank is represented by 18 perspex cylinders, all of which are removable, together with four permanently fixed half-cylinders. The copper cylinder may be inserted in place of any one of the 18 perspex cylinders, thus permitting exploration of the heat transfer rate in any row of cylinders. Alternatively all the cylinders may be removed and the copper cylinder inserted in the middle position in the front row, permitting the heat transfer to an isolated cylinder to be investigated.

66

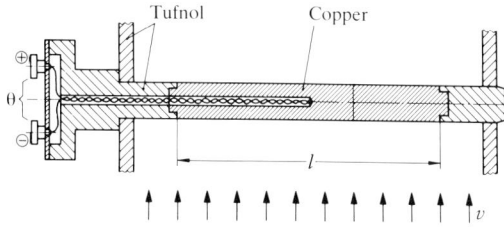

Fig. 3.6.3 Cross-flow heat exchanger: cylindrical copper element

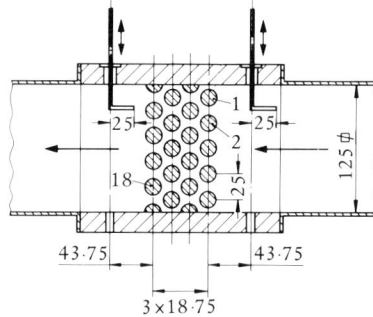

Fig. 3.6.4 Dimensions of working section

Knowing the mass m, the specific heat c and the surface area A of the copper cylinder, it is possible to calculate the heat transfer coefficient directly from a knowledge of the cooling curve. (A once-for-all correction factor for end-effects, arising from heat loss from the copper cylinder into the plastic extension and subsequent heat transfer to the fluid stream, is made by comparing the heat transfer coefficient obtained from several cylinders of different length and extrapolating to infinite length. This leads to an assessment of the "effective" surface area of the cylinder.)

The rate at which the internal energy of the copper cylinder decreases may be calculated on the basis of equation (3.4):

$$\frac{dQ}{dt} = \dot{Q} = mc\frac{dT_c}{dt} = mc\dot{T}_c \tag{3.16}$$

where m = mass of copper cylinder, c = specific heat, A = surface area and T_c = cylinder temperature.

The rate of heat loss may be equated with the rate of heat transfer to the airstream, calculated on the basis of equation (3.5):

$$\frac{dQ}{dt} = -\alpha A(T_c - T_a) \tag{3.17}$$

where α = heat transfer coefficient and T_a = air temperature. The minus sign signifies that the temperature is falling.

67

Combining (3.16) and (3.17):

$$mc\frac{dT_c}{dt} = -\alpha A(T_c - T_a) \tag{3.18}$$

Rearranging and integrating:

$$\int_{T_{cl}}^{T_c} \frac{dT_c}{(T_c - T_a)} = -\int_0^t \frac{\alpha A}{mc} dt$$

where T_{cl} = temperature of cylinder at time $t = 0$:

$$\log_e\left(\frac{T_c - T_a}{T_{cl} - T_a}\right) = \frac{-t}{\tau} \tag{3.19}$$

where the time-constant $\tau = \dfrac{mc}{\alpha A}$

Equation (3.19) may be written:

$$(T_c - T_a) = (T_{cl} - T_a)e^{-t/\tau} \tag{3.20}$$

This is a statement of Newton's Law of Cooling, and gives a curve of the form sketched in Fig. 3.6.5. The temperature of the cylinder approaches that of the airstream asymptotically, the slope of the curve being directly proportional to the temperature difference.

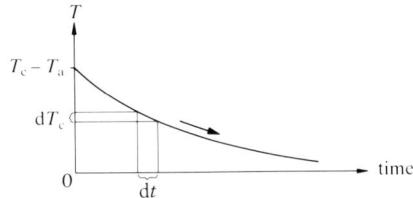

Fig. 3.6.5 Cooling curve

The pen recorder connected to the cylinder thermocouple produces curves of exactly this form, and we may use equations (3.19) or (3.20) to calculate α from these curves by the following procedure. Choose some point near the start of the cooling curve as representing the co-ordinates $t = 0$, $T_c = T_{cl}$. Choose a second point further down the curve with co-ordinates t and T_c. T_a is known, and the equation may therefore be solved for τ, which is itself a function of α and of other known quantities. Equation (3.19) may be expanded and rearranged for this purpose:

$$\alpha = \frac{-mc}{At}\log_e\left(\frac{T_c - T_a}{T_{cl} - T_a}\right) \tag{3.21}$$

It has been assumed in the above calculation that α is independent of temperature, i.e. that the rate of heat transfer is directly proportional to the temperature difference between cylinder and airstream. This assumption may be tested by calculating α for a number of points on a cooling curve.

3.6.1 Measurements and Calculations

In order to establish the effective velocity of the air passing the element it is necessary to measure the velocity upstream of the tube bank or single element and then to apply a correction for the blockage presented by the tube or tubes.

The stagnation pressure or dynamic head upstream is measured by a total head tube in conjunction with a pressure tapping in the walls of the duct, and the velocity v_∞ is calculated from:

$$v_\infty = 75 \cdot 04 \sqrt{\frac{hT}{p_0}}$$

where h = stagnation pressure, mmH$_2$O, T = absolute temperature and p_0 = barometric pressure, N/m^2 (see [11], Appendix 1, for derivation).

It is usual, when calculating the effective velocity through a bank of tubes, to base this on the minimum flow area. When all the tubes are present, this minimum area occurs in a transverse plane including a row of five tubes and, since the tubes have a diameter of 1·25 cm and the width of the working section is 12·5 cm, the effective area is one-half that of the working section and we may write in this case $v = 2v_\infty$. When a single element is being studied in isolation, the minimum flow area is 9/10ths of the full working section area, and we may write: $v = \frac{10}{9}v_\infty$.

The Nusselt and Reynolds Numbers are calculated as follows:

$$(Nu) = \frac{\alpha d}{\lambda}$$

$$(Re) = \frac{vd}{v}.$$

It is good practice to measure the velocity upstream of the tubes in any given test before taking the cooling curve, and then to withdraw the total head tube so that the heat transfer is not affected by the additional turbulence associated with the wake of the tube.

Fig. 3.6.6 shows a typical cooling curve illustrating the measurement of α, and a typical set of results obtained with the authors' apparatus and referring to this curve is analysed below:

Cylinder data:

$$d = 0 \cdot 012\ 42 \text{ m}$$
$$m = 0 \cdot 1093 \text{ kg}$$
$$c = 380 \text{ J/kg K}$$
$$A = 0 \cdot 004\ 04 \text{ m}^2$$

Cylinder in centre of fourth row

Flow conditions:

$$T_a = 291\text{K}$$
$$h_0 = 761 \text{ mmHg}$$
$$p_0 = 101\ 500 \text{ N/m}^2$$
$$h = 14 \cdot 45 \text{ mmH}_2\text{O}$$
$$v_\infty = 15 \cdot 4 \text{ m/s}$$
$$v = 30 \cdot 8 \text{ m/s}$$

69

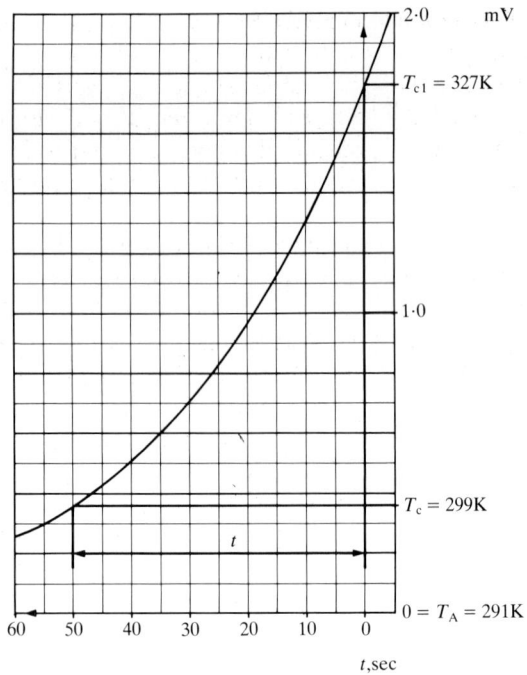

Fig. 3.6.6 Pen recording of cooling curve showing derivation of α

Cooling curve:

$$T_{c1} = 327\text{K}$$
$$T_c = 299\text{K}$$
$$t = 50 \text{ s}$$

From equation (3.21):

$$\alpha = \frac{-0 \cdot 1093 \times 380}{0 \cdot 004\,04 \times 50} \log_e \left(\frac{8}{36}\right)$$
$$= 309 \text{ W/m}^2 \text{ K}.$$

For air at inlet conditions:

$$\lambda = 0 \cdot 0259$$
$$v = 15 \cdot 38 \times 10^{-6}$$
$$(Pr) = 0 \cdot 71$$
$$\text{giving } (Re) = \frac{30 \cdot 8 \times 0 \cdot 012\,42}{15 \cdot 38 \times 10^{-6}} = 24\,900$$
$$(Nu) = \frac{309 \times 0 \cdot 012\,42}{0 \cdot 0259} = 148.$$

3.6.2 Discussion of Results

(a) *Isolated Tube*

The boundary layer around the forward surface of the tube will be laminar as a consequence of the fairly low Reynolds Number [11, p. 85]. The Nusselt Number, which incorporates the tube diameter as the characteristic length, ranges from about 20 to 90; this implies (p. 39) that the same rate of heat transfer would be achieved if the cylinder were surrounded by a layer of air at rest of thickness in the range $d/20$ to $d/90$ (0·6 to 0·13 mm), and that outside this layer the heat were conducted without further temperature difference. The thickness of the actual boundary layer must be of similar magnitude.

The Nusselt Number increases approximately with $\sqrt{(Re)}$ and hence with the square root of the velocity. We know from investigations in fluid mechanics [11] that the thickness of the laminar boundary layer varies with $1/\sqrt{(Re)}$; the relation between (Nu) and (Re) is thus consistent with the mechanism of heat transfer described above. Strictly speaking this description applies only to conditions in the boundary layer, but in the wake also the intensity of turbulence increases with Reynolds Number, with a consequent tendency for the thickness of the layer of air in contact with the cylinder in which laminar conditions apply to decrease.

A comparison of the results of an experiment with the authors' apparatus with a summary of data taken from the literature, Fig. 3.6.7, shows that our results indicate heat transfer rates slightly higher than those predicted in the literature. This suggests a level of turbulence in our apparatus higher than typical for experiments of this kind. Our results could be represented by a curve of the form:

$$(Nu) = 0.25(Re)^{0.6}$$

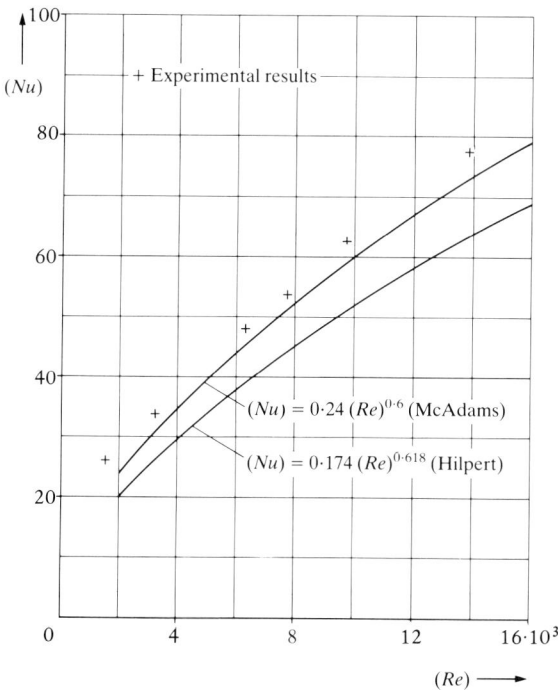

Fig. 3.6.7 Relation between Nusselt and Reynolds Numbers for isolated cylinder in cross flow

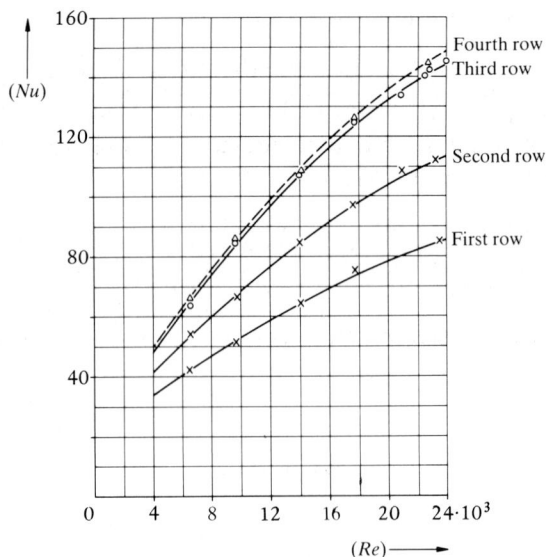

Fig. 3.6.8 Relation between Nusselt and Reynolds Numbers for tube bank

(b) *Tube Bank*

The authors' results, Fig. 3.6.8, show that in the first row of the bank the rate of heat transfer is substantially less than is the case for the isolated cylinder: $(Re) = 14000$ corresponds to $(Nu) = 65$ as compared with $(Nu) = 77$ for the isolated cylinder. This may be simply explained by the circumstance that (Re) for the tube bank is calculated with reference to the air velocity in the plane formed by the axes of the tube row. (Re) in the transverse plane corresponding to the forward surface of the tubes is clearly much lower, and the corresponding boundary layer will be thicker.

The conditions in the second and latter rows are different: the air has been accelerated to a velocity corresponding to the nominal Reynolds Number in the course of its flow through the passages between the tubes of the first row, which act as nozzles. This velocity is maintained up to the point of impact with the tubes in the second row, the heat transfer conditions for which are substantially identical with those for the isolated tube at the same Reynolds Number. The relation between (Nu) and (Re) for the second row is very close to that for the isolated cylinder.

The tubes in the second row are swept by jets of air emerging from the spaces between the tubes in the first row. The tubes in the third row are swept by the jets emerging between the tubes in the second row; these have the further feature that each jet is formed of two streams joining at an angle of about 90° and emerging from the diagonal spaces between tubes in the first and second rows. As a consequence the turbulence level in the third row is substantially higher than that in the second, and this is reflected in the higher rate of heat transfer. The flow conditions remain almost constant in the later rows, and the heat transfer rate for the fourth row is only very slightly greater than that for the third.

72

3.6.3 Further Experiments and Questions

(a) Remove the tubes in the second and fourth rows and measure the heat transfer coefficient for tubes in the remaining rows.

(b) Calculate the constants k_1 and k_2 in equation (3.8) for each row of tubes by plotting the results on logarithmic paper.

(c) Qualitative experiments on the influence of turbulence on heat transfer by fitting screens at the air inlet. If a hot-wire anemometer is available, the actual level of turbulence may be measured.

(d) We have assumed that temperature gradients in the copper cylinder are negligible. Is this justified? (Table 3.1.1, p. 36.)

3.7 Experiment 3: Boiling Heat Transfer

The conventional technique for studying boiling heat transfer in the laboratory is to employ an electrically heated resistance element, usually in the form of a single wire. To keep the power input within reasonable limits, the diameter of the wire is usually small and its radius of curvature very much less than that of the heating elements in a full-scale boiler or evaporator. This introduces a scale effect of unknown magnitude. A further disadvantage of this method, already referred to, is that it does not permit study of the important unstable regime of mixed nucleate and film boiling.

For these reasons, the authors developed an apparatus, illustrated in Fig. 3.7.1 and 3.7.2, that makes use of a transient method analogous to that used for the last experiment. A solid copper cylinder of diameter 100 mm and length 35 mm is heated to a temperature of about 300°C and plunged in boiling water. Boiling takes place at the cylinder surface and the accompanying cooling curve is plotted, using thermocouples and a pen recorder. The authors did not find it practicable to raise the initial temperature sufficiently to reach the pure film-boiling regime, but in the course of cooling the copper cylinder traverses the other three boiling regimes and passes through the point of peak heat flow, Fig. 3.2.16.

The entire cooling process occupies less than one minute and the thermal capacity of the copper cylinder is therefore sufficient to support the very large rates of heat transfer associated with the comparatively large surface in contact with the water. Pure copper is the best commercially practicable material for the cylinder, since its thermal conductivity is higher than that of any other metal with the exception of silver, while its specific heat per unit volume is also high. Nevertheless, the maximum rate of heat transmission from the surface of the cylinder is so great that internal temperature gradients cannot be neglected. To deal with this factor, two thermocouples of copper–constantan are employed, one embedded flush with the cylinder surface at the centre of one end-face which registers the surface temperature T_e. The second thermocouple is also embedded on the cylinder axis but at a distance from the centre of the mass at which, on the assumption of a uniform rate of heat transfer over the whole external surface of the cylinder, it will register the true mean temperature T_m of the cylinder. The e.m.f.s from both thermocouples are recorded simultaneously on a two-pen recorder having a traversing speed of up to 10 mm/s. It will be apparent

73

Fig. 3.7.1. Boiling heat transfer apparatus

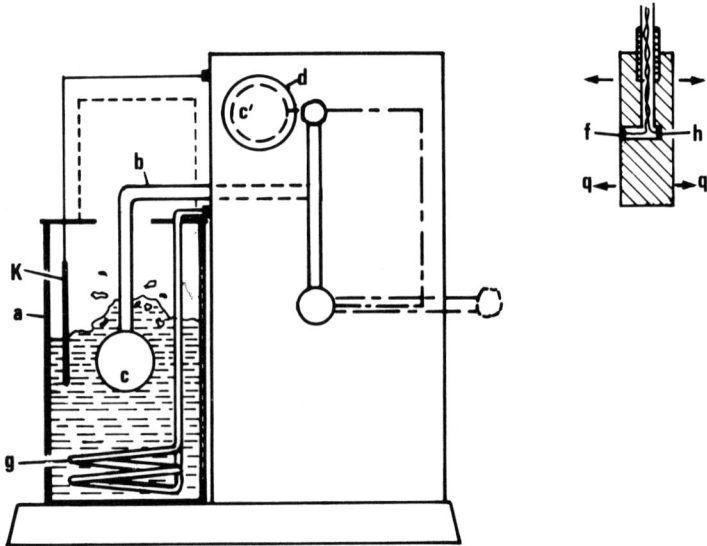

Fig. 3.7.2 Schematic arrangement of boiling heat transfer apparatus

that, if the thermal capacity and surface area of the cylinder are known, we may calculate the rate of heat transfer corresponding to any value of T_e by drawing a tangent to the curve of T_m and thus determining the rate of cooling.

The cylinder is heated to its maximum temperature by means of electrical heaters placed in contact with its end surfaces and then plunged into a cylindrical Pyrex vessel containing water which has previously been brought to the boil by an immersion heater. This heater is switched off immediately before the cylinder is plunged into the water. Arrangements are made to perform this operation without danger and to avoid splashing from boiling water. The boiling process may be clearly observed and photographed through the walls of the vessel.

3.7.1 Measurements and Calculations

It is advisable that the student should familiarize himself with the operation of the pen recorder by employing it to record the heating-up process of the copper cylinder, using a slower rate of paper feed than that appropriate to the cooling process. It will be observed that at the comparatively moderate rate of heat transfer from the electrical heaters both thermocouples record sensibly the same temperature. The high thermal conductivity of copper, roughly 8 times that of steel, ensures that under these conditions thermal gradients in the cylinder are negligible.

When the cylinder has reached a temperature of roughly 300°C and the water in the vessel is boiling the immersion heater is switched off, the traverse rate of the pen recorder increased to 10mm/s, and the cylinder plunged into the water. At the start of the cooling process the surface temperature is in the neighbourhood of point B, Fig. 3.2.16. As cooling proceeds, we traverse the unstable regime and the boiling becomes noticeably more violent as the point of maximum heat flux is approached, while simultaneously the rate of cooling indicated by the pen recorder increases. The rate of boiling then declines as we traverse the nucleate boiling regime. Eventually pool boiling sets in and the temperature of the cylinder tends asymptotically to that of the boiling water. It is advisable to reduce the paper traverse speed when the cylinder temperature has fallen to within about 5°C of the water temperature so as to reduce the length of the recording to reasonable proportions and make it easier to draw accurate tangents to the cooling curve in this region.

Fig. 3.7.3 shows a typical pair of cooling curves. It is important to note that with the usual design of two-pen recorder there is an offset on the time axis between the two traces. This offset, designated K in Fig. 3.7.3, must be taken into account when determining the simultaneous values of T_m and T_e. It will be observed that at the maximum heat flux the mean temperature of the cylinder is about 15°C higher than the surface temperature; despite the high conductivity of copper, the internal temperature gradients are appreciable.

The rate of heat transfer at any particular surface temperature is determined in accordance with the following equation:

$$\dot{q} = \frac{\dot{Q}}{A} = \frac{1}{A} \cdot mc \frac{dT_m}{dt}$$

(3.22)

where A = surface area of cylinder, m = mass of cylinder, c = specific heat of copper, and dT_m/dt = rate of cooling of cylinder.

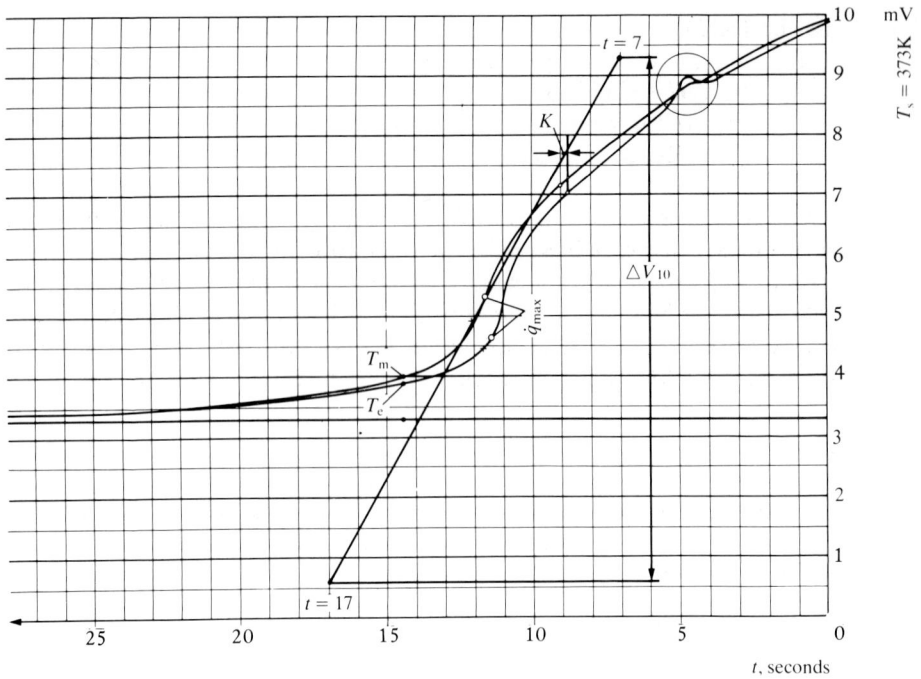

Fig. 3.7.3 Pen recording of cooling curve, boiling heat transfer

The rate of cooling is determined by carefully drawing a tangent to the curve of T_m and extending this sufficiently to cover a period of 10 seconds. This enables us to read accurately the change in thermocouple e.m.f. in 10 s, ΔV_{10}. We now determine the change in thermocouple e.m.f., Δe_{10}, corresponding to a change in temperature of 10°C from Table 2.8.4, choosing two temperatures from the table spanning the temperature T_m in which we are interested. The non-linearity of the thermocouple output is negligible over a span of 10 degrees. We may then write:

$$\frac{dT_m}{dt} = \frac{\Delta V_{10}}{\Delta e_{10}}$$

Substituting this value in equation (3.18) we may calculate $\dot{q}$.

For the authors' apparatus the following values applied:

$$m = 2\cdot45\,\text{kg}$$
$$c = 419\,\text{J/kgK}$$
$$A = 2 \times \pi \times 0\cdot05^2 + \pi \times 0\cdot1 \times 0\cdot035 = 2\cdot67 \times 10^{-2}\,\text{m}^2$$

For the point identified in Fig. 3.7.3:

$$\Delta V_{10} = 8\cdot6\,\text{mV}$$
$$T_m = 406\,\text{K}$$
$$\Delta e_{10} = 0\cdot488\,\text{mV}$$
$$\frac{dT_m}{dt} = \frac{8\cdot6}{0\cdot488} = 17\cdot6\,\text{K/s}$$

76

From equation (3.22):

$$\dot{q} = \frac{1}{2 \cdot 67 \times 10^{-2}} 2 \cdot 45 \times 419 \times 17 \cdot 6$$
$$= 677\,000\,\text{W/m}^2$$
$$T_e = 393\text{K}$$

To produce a complete curve of heat flow rate $\dot{q}$ against temperature difference $T_e - T_s$, read off a number of values of the e.m.f. indicated by the thermocouple at the cylinder surface, determine the corresponding value of T_e from Table 2.8.4, draw tangents at the corresponding points on the curve of T_m and determine values of $\dot{q}$ in accordance with the procedure described above. Fig. 3.7.4(a) shows a plot of $\dot{q}$ against $T_e - T_s$, while Fig. 3.7.4(b) shows the same data to a double-logarithmic scale.

3.7.2 Discussion of Results

The curves of Fig. 3.7.4 agree generally with values reported in the literature, though these values show very considerable scatter depending upon a number of factors, among which may be listed the following:

(a) Material from which the boiling surface is constructed.

(b) Nature of surface finish and degree of roughness.

(c) Effect of freshly adsorbed air on the boiling surface.

(d) Presence of scale and other deposits.

(e) Impurities in the water.

(f) Curvature of surface and whether this is vertical or horizontal.

(g) Volume of water and degree of convective circulation.

No satisfactory unified treatment of these various influences in the technically important regime of nucleate boiling and with regard to their influence on the maximum heat flux has yet been developed. Results for the technically less important regimes of surface and film boiling are less influenced by these extraneous factors and show themselves more amenable to theoretical treatment.

The cooling curve of Fig. 3.7.3 shows an anomaly at A which has nothing to do with boiling heat transfer but is associated with a metallurgical effect in the copper of which the cylinder is made. It is a consequence of the cold working of the copper billet from which the cylinder is machined which shows itself as a discontinuity in the thermal properties of the copper at a temperature of about 230°C. The part of the cooling curve affected should be ignored.

3.7.3 Further Experiments and Questions

(a) What mass of water is vaporized in the course of an experiment?

(b) Investigation of surface or local boiling. This occurs when the temperature of the water at the start of the experiment is lower than its boiling point. Note that in these circumstances the temperature of the water increases in the course of the boiling process.

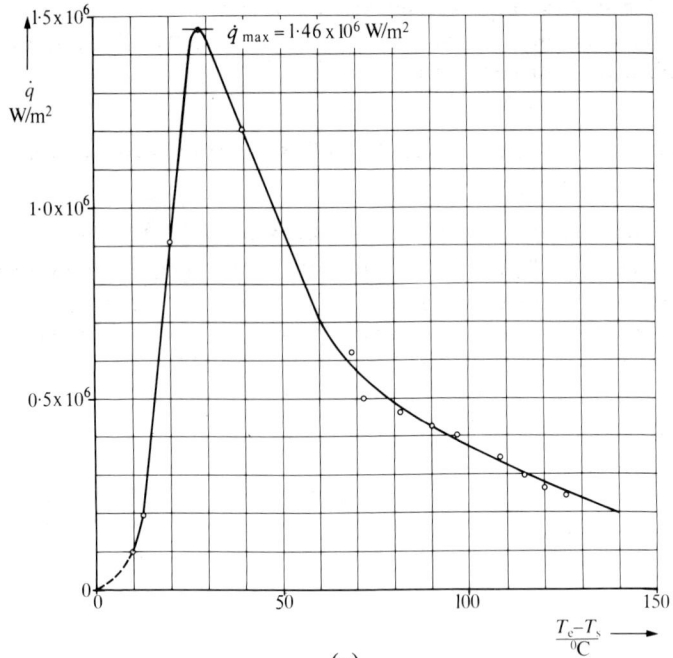

(a)

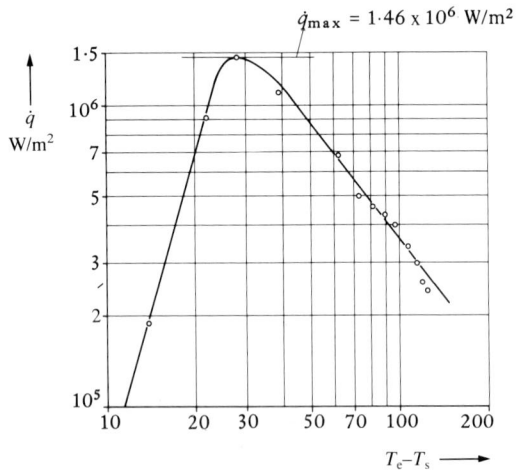

(b)

Fig. 3.7.4. Relation between rate of heat transfer and temperature difference, surface to water, boiling at atmospheric pressure:
(a) Natural scale   (b) Logarithmic scale

(c) Using cold water, confirm the thermal capacity of the copper cylinder by performing a test and equating the heat loss from the cylinder with the heat gained by a measured mass of water in the vessel.

(d) Experiments with other fluids. With some common organic fluids having boiling points at atmospheric pressure substantially lower than that of water, it is possible to explore the film-boiling regime fully within the temperature limits of the apparatus. Great care must be taken, of course, to ensure that neither the liquid nor its vapour are flammable, toxic or corrosive.

3.8   Experiment 4: Combined Heat Transfer by Free Convection and Radiation

The authors' apparatus is shown in Fig. 3.8.1 and 3.8.2. A cylindrical copper element of diameter 6·52 mm and length 160 mm is suspended in a vessel that may be either evacuated or pressurized. The element, which may be heated electrically, has a matt black surface. The dimensions of the vessel are large enough to permit substantially free convection from the element. The heat input to the element may range up to about 10 watts corresponding to a maximum surface temperature of 200°C. With this very small heat input, heating of the containing vessel is negligible and the temperature of the "atmosphere" in which the element is suspended may be taken as equal to that of the vessel and is measured by a thermocouple in the vessel wall. The temperature of the surface of the heated cylinder is measured by a nickel–chromium/nickel–aluminium thermocouple brazed to its surface.

The vessel may be charged with air or another gas up to a pressure of 2 atmospheres absolute, and the pressure may also be reduced to a minimum of about 0·03 mmHg by means of a vacuum pump. The pressure in the vessel is measured either by a mercury U-tube in conjunction with a reading of the barometer or by a McLeod gauge when the pressure is less than about 150 mmHg.

Within the range of gas pressures and temperatures with which we are concerned the rates of heat transmission from the element by radiation and by free convection

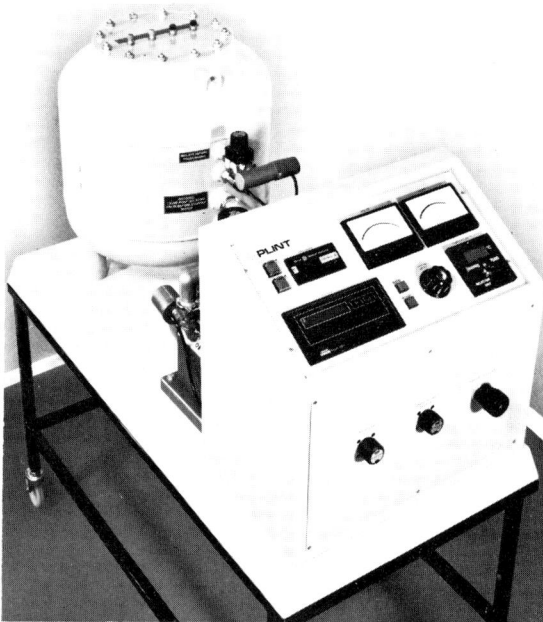

Fig. 3.8.1.   Natural convection and radiation apparatus

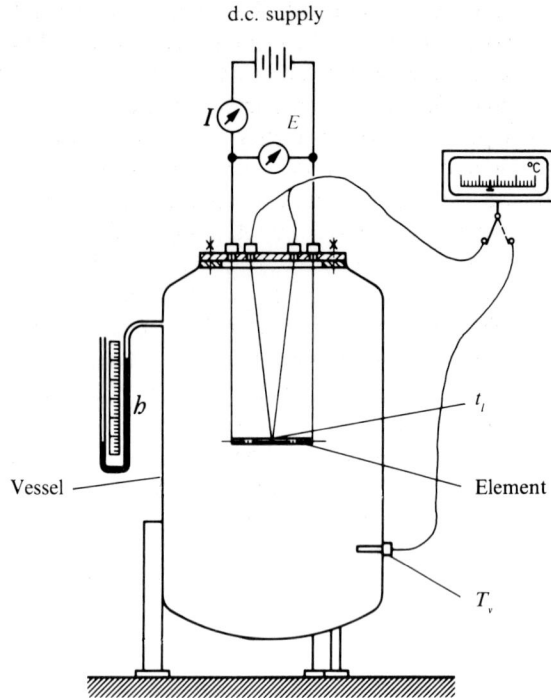

Fig. 3.8.2 Schematic arrangement, natural convection and radiation apparatus

are comparable, and the principal experimental difficulty lies in the separation of these two effects. While in a perfect vacuum heat transfer could only take place by radiation, the nature of the convective process is such that even at the quite low pressure attainable in our apparatus the convective losses are still substantial (see p. 57). The experimental method thus involves measuring the heat losses down to the minimum attainable pressure and then extrapolating to zero pressure to determine the radiation losses. A problem arises in establishing the correct method of making this extrapolation.

3.8.1 Measurements and Calculations

It is recommended that for simplicity the first experiments should be carried out using air as the surrounding gas; subsequently work may be done with other gases. The method is first to charge the vessel with air up to a pressure of about 1 atmosphere gauge from a compressed-air supply. The absolute pressure in the vessel is then determined by reading the mercury U-tube and adding the barometric pressure. The power supply to the element is switched on and the rheostat adjusted to give an energy input of about 4W. The temperatures t_e of the element and t_v of the vessel are observed every two or three minutes until t_c becomes steady. It is a feature of experiments in the field of free convection that a substantial lapse of time is necessary to achieve stable conditions. This is because the rates of heat transfer are generally very low compared with the minimum attainable thermal capacity of the elements from which the heat is emitted. The authors investigated a number of different methods of construction of their cylindrical element but were unable to produce a design that would give stable conditions in less than about 10 minutes. This is a typical experimental difficulty: some experiments in their very nature take a long time to perform. (If a pen recorder is available it may be used to advantage; the thermocouples are connected to it and, with a suitable chart speed, the final equilibrium temperature may be estimated without unduly delaying the readings.)

The pressure p, the e.m.f. V, and current I supplied to the element and the temperatures t_e and t_v are recorded, together with the barometer reading. The pressure in the vessel is then reduced and a further reading is taken when stable conditions have been achieved, the power supplied to the element being held constant. A series of results are taken with progressive reduction of the pressure in the vessel to atmospheric. The vacuum pump is then started up, the change-over valves set appropriately, and further results are taken with progressively lower pressures in the vessel, still with constant power input. It will be observed that with declining pressure the temperature of the element increases and a larger proportion of the heat loss takes the form of radiation. If the sealing is adequate an ultimate pressure of about 0·03 mmHg should be attainable, and a final reading is taken at this pressure.

It is necessary to apply various corrections to determine the true heat input to the element. In the authors' apparatus losses due to the electrical resistance of the leads supplying the element amounted to 4 percent of the energy supplied so that the true heat input to the element was given by:

$$\dot{Q}_i = 0.96 \text{ V.I}$$

A small part of the heat supplied to the element is lost through conduction along the leads supplying the power to the element (which also support it) and along the thermocouple wires. This gives rise to a loss:

$$Q_1 = 0.001\ 68(t_e - t_v)$$

The net heat loss due to convection and radiation is thus given by:

$$\dot{Q} = 0.96 \text{ V.I.} - 0.001\ 68)t_e - t_v)$$

The surface area of the element including the end surfaces amounted in the authors' apparatus to:

$$A = 0.003\ 33\,\text{m}^2$$

allowing us to calculate:

$$\dot{q} = \dot{Q}/A$$

Before we can determine the laws governing heat loss from the element due to free convection we must establish the radiation losses and this involves a determination of the emissivity ε. An obvious solution would appear to be to plot the temperature difference $t_e - t_v$ against the absolute pressure and extrapolate this to zero pressure. However, Fig. 3.8.3, plotted from a typical set of observations, shows that the shape of the curve is such that no meaningful extrapolation is possible. The method of dealing with this problem is explained below.

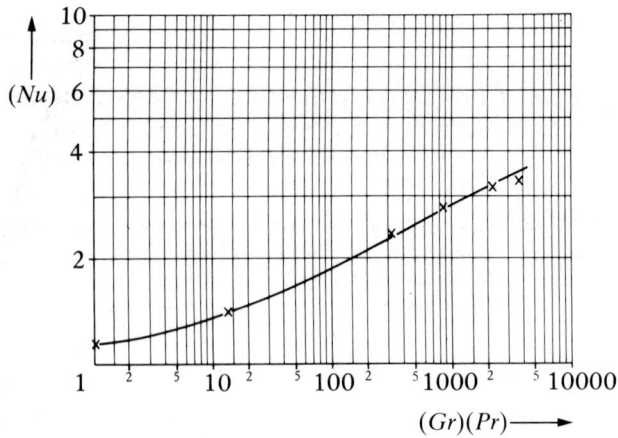

Fig. 3.8.3 Relation between Nusselt and product of Grashof and Prandtl Numbers, horizontal cylinder in air

Table 3.8.1 Experiment 4: Combined Heat Transfer by Convection and Radiation: Observations

Point	V volt	I amp	P_{abs} mmHg	t_e °C	t_v °C
1	5·90	0·59	1616	74	28
2	5·95	0·59	1229	75·5	28
3	5·95	0·59	755	80	28
4	5·95	0·59	449	85	29
5	5·95	0·59	85	97	27
6	5·95	0·59	23	103	28
7	5·95	0·59	0·09	121	28
8	5·95	0·59	0·03	125	28

Ambient temperature   20°C
Barometer   737 mmHg

Table 3.8.1 shows a set of observations, and points 7 and 8 are analysed below. For point 8:

$$\dot{Q} = 0.96 \times 0.59 \times 5.90 - 0.001\,68 \times 97$$
$$= 3.207 \text{ W}$$
$$\dot{q} = \frac{3.207}{0.003\,33} = 963 \text{ W/m}^2$$

Even at this low pressure, convective heat transfer is responsible for about 10 percent of the heat loss from the element. An estimate is made of the convection loss using the empirical equation (3.15).

At the mean temperature of the element and its atmosphere, 76·5°C, 349·5K, the mean free path (p. 58) is given by:

$$\lambda = 159 \times 10^{-9} \cdot \frac{349\cdot5}{0\cdot03} = 1\cdot85 \times 10^{-3} \, \text{m}$$

$$(Kn) = \frac{1\cdot85 \times 10^{-3}}{0\cdot006\,52} = 0\cdot284$$

The air density $\rho = 40 \times 10^{-6} \, \text{kg/m}^3$

The dynamic viscosity of gases is effectively independent of pressure and dependent only on temperature.

For air at 349·5K:

$$\mu = 2\cdot0 \times 10^{-5} \, \text{kg/ms}$$

whence $\qquad\qquad v = \mu/\rho = 0\cdot50 \, \text{m}^2/\text{s}$

The Grashof Number (p. 40), noting that $\beta = 1/T = 1/349\cdot5$, is:

$$(Gr) = \frac{9\cdot81 \times (1/349\cdot5) \times 97 \times 0\cdot006\,52^3}{0\cdot50^2} = 3\cdot0 \times 10^{-6}$$

Taking $(Pr) = 0\cdot71$, $\lambda = 1\cdot4$ and $a = 0\cdot96$, equation (3.15) yields:

$$(Nu) = 0\cdot278$$

Taking $\qquad\qquad \lambda = 0\cdot0291 \, \text{W/mK},$

$$\alpha = \frac{(Nu)\lambda}{d}$$

$$= \frac{0\cdot278 \times 0\cdot0291}{0\cdot006\,52}$$

$$= 1\cdot24 \, \text{W/m}^2\text{K}$$

Then convective heat loss

$$= 1\cdot24 \times (125 - 28) \times 0\cdot003\,33$$
$$= 0\cdot400 \, \text{W}$$

whence radiation loss:

$$3\cdot207 - 0\cdot400 = 2\cdot807 \, \text{W}$$

Finally, the emissivity is calculated from equation (3.14):

$$\varepsilon = \frac{2\cdot807}{0\cdot0033 \times 5\cdot77(3\cdot98^4 - 3\cdot01^4)}$$
$$= 0\cdot87.$$

A similar calculation for point 7 yields $\varepsilon = 0\cdot90$, and the agreement may be regarded as satisfactory.

For the remaining test points the heat transfer due to radiation is calculated using a mean value $\varepsilon = 0\cdot885$ and subtracted from the total heat transferred to the surroundings to give the convective heat loss. The corresponding Nusselt, Grashof and Prandtl Numbers are calculated and the results are plotted in Fig. 3.8.3. This curve is shown dotted in Fig. 3.2.14 and indicates good agreement with accepted values.

3.8.2 Discussion of Results

The experiment confirms the surprising prediction of Section 3.4 that the heat transfer due to free convection remains substantial even at very low pressures. For our lowest experimental point, corresponding to a pressure of $0\cdot03$ mmHg, convective heat transfer is still responsible for some 12 percent of the total heat loss. At this pressure, $(Kn) = 0\cdot284$ corresponding to a mean free path of the air molecules of $\lambda = (Kn) \cdot d = 0\cdot284 \times 6\cdot52 = 1\cdot85$ mm. There is thus no boundary layer in the accepted sense of the word. Equation (3.15) suggests that for the same Grashof Number but with $(Kn) = 0$ the heat transfer would be some 15 percent greater. The decline in heat transfer rate with increasing Knudsen Number (i.e. with falling pressure and increasing mean free path) is a consequence of the decline in convective flow and the increasing dependence of the heat transfer process on pure conduction.

Fig. 3.8.3 is a striking illustration of the power of the method of dimensional analysis in the presentation of experimental data. The curve is drawn on the basis of a series of tests in which only the total power input to the heated element remained constant. The other independent variables, pressure, temperature and convective heat loss, all varied from one point to the next. Despite this, the curve is of a perfectly general nature, and this may be demonstrated by taking a number of sets of readings, each at a constant pressure, in which the power input to the element is varied and the corresponding temperature observed.

These results may be plotted as a series of curves of convective heat loss against temperature, all of which may be shown to coincide with the curve of Fig. 3.8.3 when plotted non-dimensionally as (Nu) against $(Gr).(Pr)$.

3.8.3 Further Experiments and Questions

(a) Experiments with other gases (carbon dioxide, helium, etc.).

N.B. If it is proposed to use flammable or explosive gases such as hydrogen very special precautions must be taken during the process of purging and charging the vessel. The manufacturer's instructions should be followed with care.

(b) Confirmation of the Stefan–Boltzmann equation by repeating the experiment for a range of different energy inputs and hence of element temperatures.

(c) Experiments with elements of different diameter and with different surface treatments.

(d) Apply the results of the experiment to estimate the heat loss from an ordinary hot water heating radiator. What is the approximate division of heat transfer between radiation and free convection at a water temperature of 80°C? (see Fig. 3.2.14).

4

The First Law of Thermodynamics

4.1 The Steady-flow Energy Equation

In Section 2.5 we have stated the First Law of Thermodynamics in terms of its application to closed or non-flow systems. For such systems it represents the balance between work, heat and internal energy. Most engineering applications of thermodynamics, however, are concerned with plant or machinery through which a continuous flow of fluids takes place. These fluids may be fuel, air, water, or products of combustion; they may also include working fluids such as steam or compressed air. Such devices are known as open systems. The First Law of Thermodynamics is applied to such steady-flow processes in terms of the steady-flow energy equation (often abbreviated to s.f.e.e.), which represents an energy balance analogous to that developed for non-flow processes in Chapter 2.

Fig. 4.1.1 shows a simple example of a machine the operation of which may be described in terms of a steady-flow process: a compressed-air motor. The system boundary is shown as a dotted line; it differs from the corresponding boundary applicable to a non-flow process in that compressed air, exhaust air and work flow continuously across the system boundary. It may, of course, be the case that cyclic processes are taking place within the machine, but this does not affect the analysis; the rate of flow of fluid and work into and out of the system is regarded as constant, and the total of energy and working fluid contained within the system boundary is also taken to be effectively constant.

In the case of a closed system the physical configuration of the system boundary usually changes with time to follow changes in the volume of the constant mass of working fluid contained within the system. In the case of open systems, however, the system boundaries remain unchanged with time.

Consider the operation of the compressed-air motor. It will be assumed that the motor has been running for some time, so that conditions of air flow, power output and temperature are steady. The motor may be either warmer or cooler than its surroundings, with the consequence that heat Q_{12} may flow from the surroundings to the motor or vice versa. (Flow to the motor is regarded as positive.)

In order to apply the First Law to this situation we first convert the open system to a closed one by extending the system boundary from the compressed air inlet of the motor to enclose a mass m of compressed air corresponding to the amount entering the machine during one second, Fig. 4.1.1(b). Similarly, we extend the system

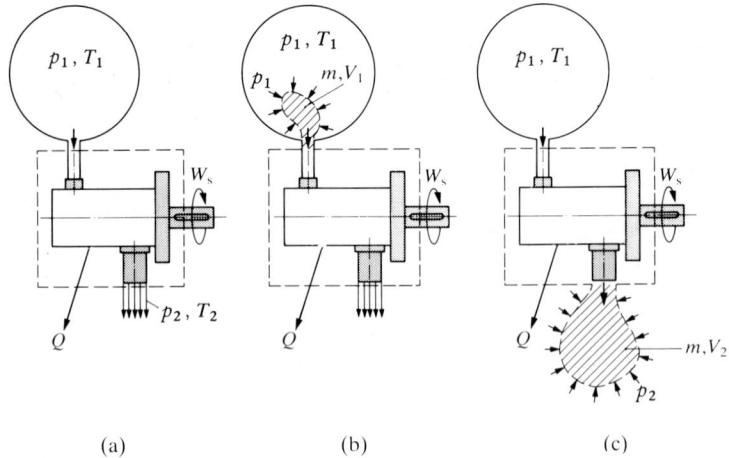

Fig. 4.1.1 Application of the steady-flow energy equation to a compressed air motor

boundary at the motor exhaust to enclose the same mass m of air discharged by the motor in one second, Fig. 4.1.1(c).

We can now apply the First Law in the form of equation (2.5) to this closed system. State 1 corresponds to Fig. 4.1.1(b), state 2 to Fig. 4.1.1(c). The change in internal energy $U_2 - U_1$ over the period of one second corresponding to the change from state 1 to state 2 is equal to the change in internal energy of the mass m of air in its passage from state 1 (at air motor inlet) to state 2 (at air motor outlet). It is evident that the internal energy of the air within the motor remains constant in time.

We now regard the system boundary as again fixed, Fig. 4.1.1(a), and consider the flow of work through the boundary in a period of one second. Besides the mechanical or shaft work W_s performed by the air motor and leaving the system, we have to consider the displacement work crossing the boundary at air inlet and exhaust. At the inlet, the compressed air entering the system conveys work $p_1 V_1$ into the system. (This may be regarded as the work performed by the contents of the compressed air receiver on the "balloon" of air entering the machine, Fig. 4.1.1(b), in the course of forcing it into the machine.) Similarly, during the expulsion of the exhaust air from the machine work $p_2 V_2$ is performed by the machine on the surroundings.

While not all authorities agree on the use of the term, the authors prefer to describe the quantities $p_1 V_1$ and $p_2 V_2$ as flow work. It is sometimes argued that flow work does not represent a genuine work transfer through the boundaries of the system, but it is only necessary to consider the case of a machine working on an incompressible working fluid, such as a high-pressure oil motor, to show that this is not the case. The power output from an oil hydraulic motor is derived entirely from the flow work of the high-pressure oil entering the machine.

We may now write equation (2.5) as follows:

$$U_1 - U_2 = W_s + p_2 V_2 - p_1 V_1 - Q_{12} \tag{4.1}$$

The equation has been rearranged to present the terms in a logical manner: the decrease in internal energy of the compressed air in its passage through the machine,

86

$U_1 - U_2$, is equated with the work leaving the machine, $W_s + p_2 V_2$, less the work entering the machine, $p_1 V_1$, less heat gained by the machine, Q_{12}. U, p and V are properties while W_2 and Q_{12} are not. Bringing both groups together, we may write:

$$(U_1 + p_1 V_1) - (U_2 + p_2 V_2) = W_s - Q_{12} \tag{4.2}$$

The quantity $(U + pV)$ is built up from properties of the working fluid, is independent of the previous history of the fluid and is therefore itself a property. It is given the name "enthalpy", H:

$$H = U + pV \tag{4.3}$$

Using this new property, we may write the First Law of Thermodynamics as applied to flow processes, the steady-flow energy equation (s.f.e.e.), as follows:

$$H_1 - H_2 = W_s - Q_{12} \tag{4.4}$$

To apply equation (4.2) to a quantity of 1 kg of working fluid we divide through by m and then write:

$$h_1 - h_2 = w_s - q_{12} \tag{4.5}$$

where h_1 and h_2 are specific enthalpies corresponding to 1 kg of air.

The power output of a machine such as the air motor may be written:

$$P_s = w_s \dot{m} = \dot{m}(h_1 - h_2) + \dot{Q}_{12} \tag{4.6}$$

where $\dot{m} =$ mass flow rate of working fluid.

The property enthalpy is of great importance in practical thermodynamic calculations, and tables and charts of enthalpy have been prepared for technically important working fluids and in particular for air and water. As is the case for all pure substances, the state may be defined by two independent properties. Thus, for example, the enthalpy of steam may be determined from the tables if one knows the pressure and specific volume.

To illustrate the importance of equation (4.6), consider a machine such as a large steam turbine in which the heat transfer to and from the surroundings $\dot{Q}$ is negligible relative to the other terms of the equation. Knowing the inlet and outlet pressure and temperature of the steam and the mass flow rate $\dot{m}$, then using this equation we can immediately make an approximate calculation of the power output. If allowance is made for bearing friction and similar parasitic losses, the estimate comes very close to the actual power output of the machine.

Throttling is an important steady-flow process. This is a process in which working fluid at high pressure is expanded, for example through a valve, without exchange of heat with the surroundings and without performing mechanical work, Fig. 4.1.2. For such a process equation (4.6) reduces to:

$$h_2 = h_1 \tag{4.7}$$

This equation enables us to determine the temperature after a throttling process if we know the pressure and temperature of the fluid before throttling and the downstream pressure. We look up the value of the enthalpy $h_1 = h_2$ and determine the value of T_2 corresponding to the pressure p_2.

A further important example of a steady-flow process is the heat exchanger, in

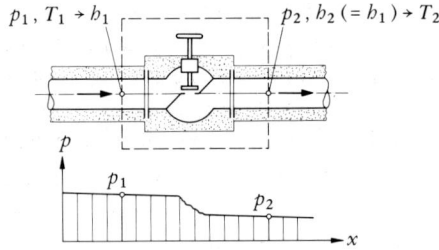

Fig. 4.1.2 The throttling process

which heat but no mechanical work is exchanged. In this case we may re-write equation (4.6):

$$\dot{m}_a(h_{a1} - h_{a2}) = \dot{m}_b(h_{b2} - h_{b1}) - Q_{12} \tag{4.8}$$

assuming that fluid a is transferring heat to fluid b and noting that as the heat exchanger generally loses heat to its surroundings Q_{12} is negative.

In the special case in which the progress of the working fluid through the system may be represented on a pressure–volume diagram of known form, Fig. 4.1.3, we may write:

$$W_s = \int_1^2 V \, dp \quad \text{or} \quad w_s = \int_1^2 v \, dp \tag{4.9}$$

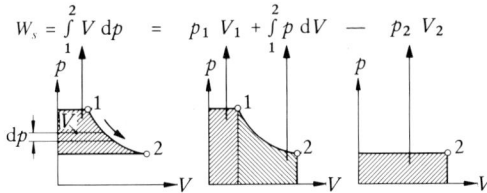

Fig. 4.1.3 Displacement work

This represents the sum of the expansive work $\int p \, dV$ and the net flow work $p_1 V_1 - p_2 V_2$.

In this discussion we have ignored terms that may appear in the steady-flow equation consequent on changes in the potential and kinetic energies of the working fluid between system inlet and system outlet. These differences are not usually of significance in the case of thermal machines; they are dealt with in Chapter 10.

4.2 Classification of Sources of Mechanical Power

If we exclude electric motors, virtually all sources of mechanical power operate by extracting energy from a working fluid. In the course of this process the pressure and temperature of the working fluid invariably falls, and in some cases its volume increases. The steady-flow energy equation may be applied in the analysis of the performance of all machines of this type. Such machines may be subdivided into two main classes: positive displacement machines and turbo machines. In positive displacement machines the working fluid exerts pressure directly upon moving members that are coupled ultimately to the output shaft of the machine; in the case of turbo machines all or some of the energy of the working fluid is converted into kinetic energy which is subsequently converted into useful work by the impact of the working fluid upon moving blades. Positive displacement machines, typified by the internal

Table 4.2.1 Classification of Power Sources

	Positive displacement machines	Turbo machines
Principle of operation	Static Cyclic operation and direct action of pressure upon pistons	Dynamic Continuous operation and energy transformation from pressure to velocity
Sealing	Sliding or close rolling contact	Labyrinth
Relative velocity fixed and moving parts	Typically up to 25 m/s; limited by friction and wear	Typically up to 500 m/s; limited by material strength, velocity of sound
Suited for	High pressure Small volumetric flow rates Small power Low revolutions	Both low- and high-pressure Large flow rate Large power High revolutions
Special characteristics	Bulky High efficiency	Compact Moderate efficiency
Power sources	Piston engines (petrol, diesel, hot air) Rotary engines (Wankel)	Gas turbines
	Reciprocating steam engines	Steam turbines
	Air motors Hydraulic motors	Air turbines Water turbines
Reversed machines (pumps and compressors)	Piston compressors, rotating piston and sliding vane compressors	Turbo compressors, axial and radial
	Piston pumps Gear pumps Sliding vane pumps	Centrifugal and axial flow pumps

combustion engine, involve non-flow processes repeated cyclically in a cylinder closed by a reciprocating piston. Turbo machines involve steady-flow processes, and it is characteristic that compression, the addition of heat and expansion take place in separate components of the machine.

Table 4.2.1 gives a classification of the main types of power source, also of reversed machines, the characteristic of which is that both work and energy flows are in the reverse direction; mechanical work is performed on the machine, with a consequent increase in the enthalpy of the working fluid.

Note that we have included in this table a number of machines not usually associated with thermodynamic processes. Nevertheless, First Law analysis is applicable to these machines and has in some cases been used very effectively, for example in the assessment of the efficiency of large water turbines by careful measurement of the rise in temperature of the water passing through the machine.

4.3 Experiment 5: Energy Conversion in a Compressed-air Motor

The analysis of the performance of a machine developing power from a supply of compressed air is an interesting illustration of the application of the steady-flow energy equation. Fig. 4.3.1 shows a cross-section through a vane-type air motor of the kind that is employed to drive hand tools. It consists essentially of a cylindrical rotor mounted eccentrically in a cylindrical casing. Equally spaced vanes of synthetic material are carried in radial slots in the rotor and bear against the internal surface of the casing. They are maintained in contact with the casing during the high-pressure phase by means of air at supply pressure which is admitted to the bottom of the slot in which the vanes operate by way of the port d.

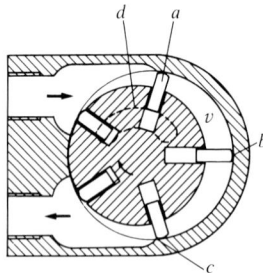

Fig. 4.3.1 Construction of vane-type air motor

The air motor operates in the following manner. In the course of the passage of a rotor vane from position a to position b air at supply pressure acts upon the vane, giving rise to a turning moment. When the vane reaches position b the succeeding vane has reached position a, sealing off the inlet port from the volume of air at approximately inlet pressure that is trapped in the space v. This volume is carried round by the rotor for a short distance until the leading vane reaches position c, when communication is opened between the trapped air and exhaust and sudden discharge of the air takes place. It will be apparent that the machine works non-expansively.

The air motor is a positive displacement machine, having a definite "swept volume"; it belongs to the class of piston machines, even though in this case the pistons take the form of sliding vanes.

Fig. 4.3.2 shows the experimental set-up used by the authors. The air motor is supplied with compressed air by way of a filter and moisture trap, a pressure regulating valve and a lubricator which injects oil into the airstream.

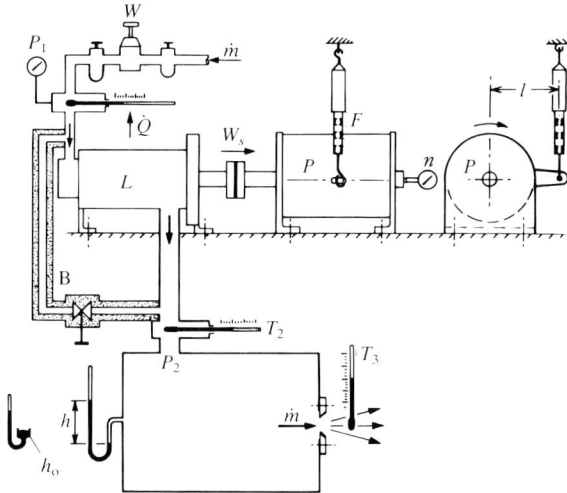

Fig. 4.3.2 Schematic arrangement of air motor test set

Inlet and exhaust air temperatures are measured by mercury-in-glass thermometers and the inlet air pressure by a pressure gauge. The rate of air flow is measured by means of a sharp-edged orifice on the exhaust side of the machine. The orifice is mounted in the wall of an airbox, the purpose of which is to eliminate pressure pulsations, and the pressure difference across the orifice is indicated by a water manometer.

The mass flow rate of air is given by:

$$\dot{m} = 0 \cdot 025 C_d . d^2 \sqrt{\frac{hh_0}{T_3}}$$

where d = orifice diameter; C_d = coefficient of discharge; h = manometer reading, mmH$_2$O; h_0 = barometric pressure, mmHg; and T_3 = air temperature at orifice, K ([11], Chapter 6).

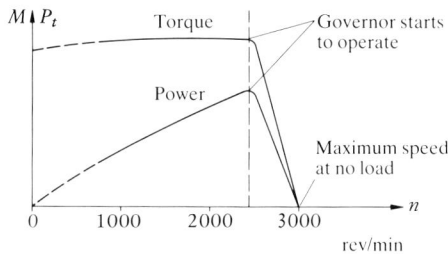

Fig. 4.3.3 Operation of air motor governor

To prevent overspeeding, the air motor is fitted with a governor which starts to come into effect at speeds above 2500 rev/min, and progressively throttles the air supply to limit the maximum speed to 3000 rev/min, as shown schematically in Fig. 4.3.3.

The power output of the air motor is absorbed by an electrical dynamometer fitted with a counter and a spring balance for torque measurement.

4.3.1 Measurements and Calculations

It is recommended that a series of measurements of power output and the other relevant data should be made over a range of speeds, maintaining the inlet air-pressure constant. It is necessary to run the machine at constant load for a sufficient time for the temperature to stabilize.

For the equipment used by the authors the following data applied:

measuring orifice diameter, $\qquad$ $d = 0.025$ m

coefficient of discharge, $\qquad$ $C_d = 0.60$

$$\dot{m} = 76.9 \times 10^{-6} \sqrt{\frac{hh_0}{T_3}}$$

$$W_t = \frac{F.n}{77.67}$$

where $F =$ spring balance reading and $n =$ rev/min.

For air at moderate temperatures and pressures, enthalpy may be regarded as effectively independent of pressure and a function of temperature only. We may then write:

$$h_1 - h_2 = 1.01(T_1 - T_2) \text{ kJ/kg},$$

taking $c_p = 1.01$ kJ/kg K as constant.

Table 4.3.1 shows a single test result, and a performance curve is shown in Fig. 4.3.4. This shows the variation in the power output W_s and in the change of enthalpy of the air passing through the machine over a range of speeds. The interval between these curves represents the heat $\dot{Q}$ exchanged between the motor and its surroundings, and it will be noted that, with the sign convention of equation (4.4), $\dot{Q}$ is positive: the motor receives heat from the surroundings. This agrees with experience; compressed-air machines invariably run cold, and icing of the exhaust system is a common problem if the compressed air is not sufficiently well dried.

The authors' machine was fitted with a by-pass incorporating a throttling valve to permit the verification of equation (4.7) as a subsidiary experiment. The air supply to the motor is shut off by means of the governor stop button, and the throttling valve in the by-pass is adjusted to give roughly the same flow rate as with the motor running at maximum speed. Inlet and outlet temperatures are observed when conditions have stabilized.

In a test by the authors the temperatures before and after the throttling valve were respectively:

$$t_1 = 30.3°C \qquad t_2 = 30.6°C.$$

Table 4.3.1 Experiment 5: Energy Conversion on a Compressed-air Motor: Observations

F Newton	15.5
n rev/min	2600
h mmH$_2$O	114
t_1 °C	293
t_2 °C	$-1\cdot2$
t_3 °C	25
p_1 bar	4·019
p_2 bar	1·019
W_s watts	519
m kg/s	0·0156
$t_1 - t_2$ deg. C	30·5
$h_1 - h_2$ kJ/hg	30·81
$\dot{m}(h_1 - h_2)$ watts	480
Q watts	-39

Ambient temperature   20°C
Barometer   757 mmHg

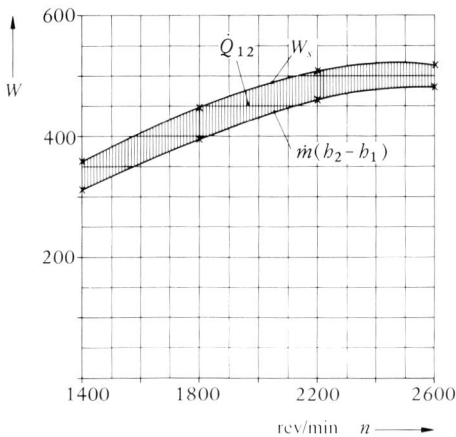

Fig. 4.3.4 Performance curve for air motor

4.3.2   Discussion of Results

It is typical of First Law analysis that the thermodynamic performance of the machine may be assessed without taking into account the exact details of the processes taking place within the machine: we observe only the flows of heat and work entering and leaving the air motor. Our analysis gives no indication of its efficiency, which can in fact be assessed on several different bases. A more detailed analysis of the performance of the machine is given in Chapter 9.

The throttling experiment confirms equation (4.7) and emphasizes the fact that the fall in temperature of the working fluid in its passage through the machine is associated with the performance of work, in this case flow work.

4.3.3   Further Experiments and Questions

It is recommended that at a later stage in the course the student should make a full analysis of the working of this machine with a view to forming a quantitative estimate of the different sources of loss. The machine may be compared with an ideal motor operating without leakage or friction and expanding the air adiabatically to atmospheric pressure and the following losses identified:

(a) Loss due to non-expansive working.

(b) Loss due to leakage.

(c) Mechanical losses.

The development engineer seeking to improve the performance of a new machine is invariably called upon to make analyses of this kind, and it is desirable that the student should gain experience in this area.

5

The Second Law of Thermodynamics

5.1 Reversibility and Availability

The usual textbook treatment of the Second Law involves a rigorous statement of the theory, and is generally somewhat abstract in nature—partly a consequence of the route by which the subject has evolved historically. This treatment may fail to make clear the very great practical importance of the Law as a guideline to the mechanical engineer concerned with the design and development of the machines* that are the main source of power in the industrialized world.

The development engineer is particularly concerned with the two concepts that form the title of this section; he seeks to approximate the processes taking place in the machine as closely as possible to reversible processes, and to maximize the availability of the heat used by the machine.

The concept of reversibility receives little attention in the field of dynamics, since most mechanical processes are either at least in theory reversible or so evidently irreversible that the possibility that they could be otherwise is not even discussed.

As a simple example, the fall of a stone from rest may clearly be regarded as in principle reversible: if we were to project the stone upwards at the appropriate velocity and from an appropriate position, it would traverse the same path at the same velocity at each point (though in the reverse direction) and would arrive at rest at its initial starting point. The resistance of the air would have a slight effect (it would introduce a degree of irreversibility) but even this could be eliminated by performing the process in a vacuum.

If, however, the stone strikes the ground the impact is quite clearly irreversible. The kinetic energy of the stone is dissipated in deforming the ground, in elastic shock-waves and in sound. Ultimately it is manifested as a minute increase in temperature of the stone and its surroundings. For the process to be reversible it would be necessary for the temperature to fall to its original value and for the corresponding energy to be released in the form of mechanical work that would propel the stone back to its original position.

It should be particularly noted that such a reversed process would in no way contravene the First Law of Thermodynamics. It is the Second Law that indicates to

---

* The term "heat engine", now less used than in the past, is generally accepted as referring only to thermodynamic systems across the boundaries of which flow only heat and work. It thus includes steam turbines but excludes internal combustion engines and gas turbines.

us which processes are reversible and which are not, and enables us to calculate the "degree" of irreversibility in cases more complex than the example given above, for which the irreversibility is evidently total.

As a further example of an essentially irreversible process consider the running down of a heavy flywheel. It is a matter of common experience that, however carefully we design the bearings, and even if we run the flywheel *in vacuo*, it will eventually come to rest. All the kinetic energy in the wheel will eventually appear as heat in the bearings and will be dissipated in the surroundings. It is clearly impossible to withdraw this heat and somehow employ it to again speed up the flywheel.

An important example of a process that can in principle be made reversible, and in practice may closely approximate to reversibility, is the resisted expansion of a gas or vapour. If we could construct a cylinder and piston of non-conducting material it would be possible to expand a gas contained in the cylinder and then to compress it to arrive back at its original condition.

In the course of expansion, work W would be performed by the "system" comprising the gas, and the internal energy of the gas would diminish by an equivalent amount. During the compression process the same amount of work would be performed on the gas to restore it to its original condition. The First Law would apply:

$$U_1 - U_2 = W,$$

where U_1, U_2 are respectively the internal energies before and after expansion. Reversible processes frequently involve the exchange of heat and work on a one-to-one basis.

In practice two sorts of loss occur: heat loss from the gas to the containing cylinder, and losses arising as a consequence of turbulence and friction within the body of the gas. The process here is quite complex. Unless the expansion takes place very slowly, velocities within the gas will be appreciable; these velocities will be generated by the expansion of the gas itself, and will result in a reduction in the force exerted by the gas on the piston. At this point no irreversibility has occurred; it would in principle be possible to recover the kinetic energy represented by the gas velocity, but in practice it is dissipated in turbulence and appears ultimately as heat.

Table 5.1.1 summarizes the types of irreversibility with which we are concerned in thermodynamic processes.

It will be clear that the process of perfecting a thermal machine involves the elimination, so far as is possible, of all the sources of irreversibility in the operation of the machine. It is obvious that we design bearings to give the minimum possible friction losses, reduce windage to a minimum, and design the internal passages of the machine so that the flow of the working fluid, air, steam, and products of combustion, through the machine takes place with the minimum possible resistance.

It is not quite so obvious that we must minimize the temperature differences associated with the transmission of heat, and it is in this area that the Second Law, together with the concept of availability, has most guidance to give us. We have already seen (p. 24) that the degree of its availability is a characteristic of a source of heat and that, in general, the higher the temperature of such a source the greater the proportion that is, at least theoretically, available for conversion into work. Any

Table 5.1.1 Sources of Irreversibility

Real process	Idealized process	Loss of actual or potential work
Relative motion of solid bodies with friction	Motion without friction	Bearing and other frictional losses
Fluid flow with friction	Flow of fluids of zero viscosity. Potential flow	Fluid friction losses
Impact	Fully elastic impact	Impact losses
Heat transmission with temperature difference	Infinitely slow heat transmission, or perfectly conducting media	Potentially available work from the heat engine
	or Substitution of a reversible heat engine with recovery of equivalent amount of work	$\dot{W} = \dot{Q}\, \dfrac{(T_1 - T_2)}{T_1}$
Flow of electric current in a conductor having resistance	Flow of current in superconductor	Resistance loss

temperature difference associated with heat transfer thus "degrades" the heat involved, reducing its potential as a source of energy of other forms.

5.2 Entropy

The foregoing discussion of irreversibility and availability gives no clue as to the basis on which these two factors may be measured. As is generally the case, it is essential, if progress is to be made in a scientific matter, to devise some entity that will give a quantitative measure of the effect concerned, in this case the loss associated with any given reduction in availability or degree of irreversibility.

Such an entity might have the following properties:

(a) In the case of reversible processes it could remain constant.

(b) In the case of an irreversible process its increase would be a direct measure of the degree of irreversibility.

(c) Processes that contravened the Second Law would imply a reduction in the entity.

Another possibility, referring to Table 5.1.1, would be to use the loss of work, or the loss of potential work, as a measure of the loss of availability or degree of irreversibility.

The evolution of thermodynamics has led to the introduction of a somewhat abstract concept, entropy, which meets the three conditions listed above. Many verbal definitions of entropy have been attempted; perhaps the most instructive is one of the oldest, attributed to Professor Sylvanus P. Thompson:

"The increase in entropy in any thermodynamic process is that quantity which, when multiplied by the lowest available temperature, gives the inevitably incurred waste."

An important advantage of the concept of entropy is that it may be applied not only to a complete system but to any substance. We may thus speak of the specific entropy (usually abbreviated to entropy) of 1 kg of a substance. Entropy is in fact a *property*, like p, V, T, U and H. It is an extensive property like V, U and H, implying that doubling the mass doubles the entropy. This implies that for a closed system the total entropy is made up of the sum of the individual entropies of the system components:

$$s_T = s_1 + s_2 + s_3 \ldots$$

In most cases we are concerned only with entropy differences, and it is thus only necessary to draw up rules for calculating entropy changes.

Consider first the case of a fully closed system that has a certain potential ability to perform work. Changes take place within the system, as a consequence of which the potentially available work is reduced. We can write the consequent increase in entropy very simply:

$$s_2 - s_1 = \Sigma \frac{W_l}{T} \tag{5.1}$$

or, in differential form:

$$ds = \frac{dW_l}{T}$$

where W_l represents the loss in potentially available work and T the lowest available or "sink" temperature, within the system.

Fig. 5.2.1 represents a closed system containing various devices that have a potential ability to perform work. It illustrates the various sources of irreversibility listed in Table 5.1.1.

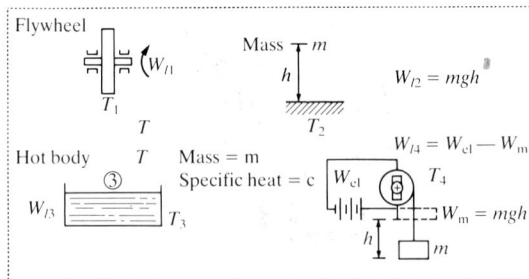

Fig. 5.2.1 Entropy changes in a closed system

First we have a rotating flywheel having a total kinetic energy W_{l1}. If the flywheel is eventually brought to rest by friction in the supporting bearings we may write the increase in entropy of the system as:

$$\Delta s_1 = \frac{W_{l1}}{T_1}$$

where T_1 is the (final) temperature of the flywheel and bearings.

98

Had the power loss been associated with fluid friction, if, for instance, the energy W_{l1} had been dissipated in stirring water at temperature T_1, the increase in entropy would have been identical.

Next consider a mass m falling through a height h and coming to rest at the lower level. The resultant increase in entropy may be written:

$$\Delta s_2 = \frac{W_{l2}}{T_2}$$

where T_2 is the final temperature of the mass and the surface on which it rests.

The system is also supposed to contain a hot body at temperature T which is quenched in a liquid, the final temperature of body and liquid being T_3.

$$\Delta s_3 = \frac{W_{l3}}{T_3}$$

Here W_{l3} is the work that could ideally have been performed by a heat engine employing the hot body as a source of heat.

Finally, the system contains an electric motor coupled to a battery. The motor performs work by lifting a weight, in the course of which some of the electrical energy drawn from the battery is dissipated as a consequence of electrical resistance and friction in the motor:

$$\Delta s_4 = \frac{W_{l4}}{T_4}$$

where T_4 is the temperature of the motor.

For the whole system:

$$s_2 - s_1 = \frac{W_{l1}}{T_1} + \frac{W_{l2}}{T_2} + \frac{W_{l3}}{T_3} + \frac{W_{l4}}{T_4}$$

As the lost work (or potential work), like the absolute temperature, is always positive, the entropy always increases.

In the case of a fall in temperature associated with conduction, Fig. 5.2.2, the work lost may be calculated by postulating a reversible engine working between the same temperature limits. We may then write:

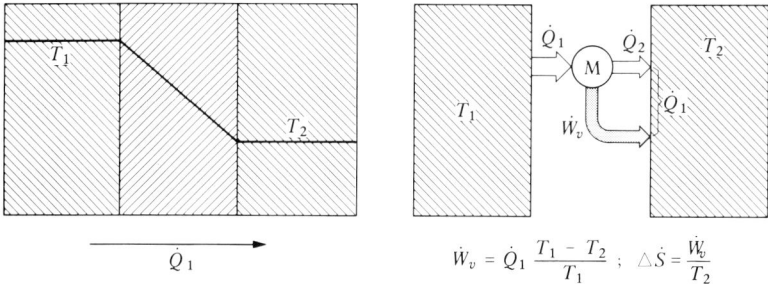

$$\dot{W}_v = \dot{Q}_1 \frac{T_1 - T_2}{T_1} \ ; \quad \Delta \dot{S} = \frac{\dot{W}_v}{T_2}$$

Fig. 5.2.2   Increase of entropy in thermal conduction

$$\dot{W}_l = \dot{Q}_1 \frac{T_1 - T_2}{T_1}$$

$$\dot{s} = \frac{\dot{W}_l}{T_2}$$

Entropy, as defined in equation (5.1), is, as mentioned earlier, a property. The proof of this is not simple, and is given in Section 10.4. It will be noted that the definition of entropy involves the absolute temperature.

So far we have been considering entropy in fully closed systems, across the boundaries of which neither heat nor work flows. In cases in which heat enters the system the increase in entropy of the system corresponding to an increment of heat dQ is given by dQ/T, where T is the temperature of the system. We may then rewrite equation (5.1):

$$ds = \frac{dQ}{T} + \frac{dW_l}{T} \qquad (5.2)$$

The flow of work either into or out of the system has no influence on the entropy. W_l refers only to losses occurring within the system.

The reader may at this point doubt the usefulness of the concept of entropy, since the losses associated with irreversibility may in simple cases be estimated by the methods summarized in Table 5.1.1. In fact, entropy plays its most important role in another field: as a property of pure substances which may be calculated in accordance with the following equation and recorded in diagrams and tables:

$$\int_1^2 T ds = \int_1^2 (dQ + dW_l) = Q_{12} + W_{l12} \qquad (5.3)$$

A pure substance in the thermodynamic sense is one that is homogeneous in chemical composition, even though its physical state may not be uniform. Examples are steam and a mixture of steam and water. Equation (5.3) implies that if we add heat to the substance and plot a curve, as in Fig. 5.2.3, the area beneath the curve will represent the sum of the heat supplied plus the equivalent of any loss of available work that has occurred in the passage of the system from state 1 to state 2.

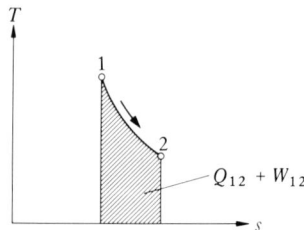

Fig. 5.2.3   Representation of heat supplied to a working fluid on the T-s diagram

The particular importance of entropy rests on the fact that it is a thermodynamic property and thus has a unique value for any state of the substance, which may be determined if any other two properties (e.g. p, T) are known. Fig. 5.2.4 is a typical example of such data, representing the relation between enthalpy and entropy for air. Other frequently used diagrams show the relation between temperature and entropy.

100

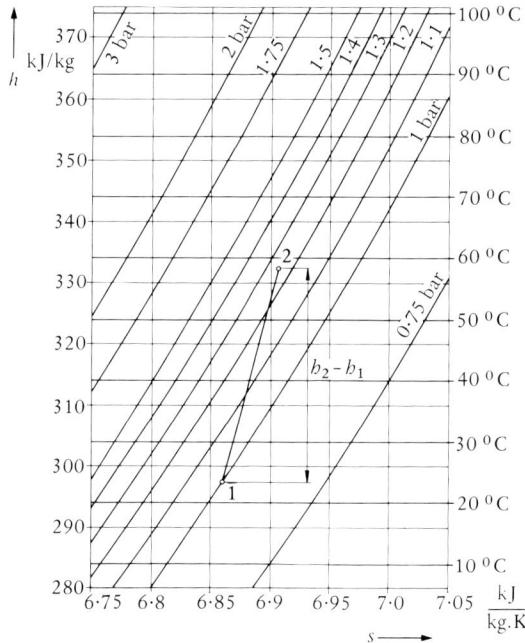

Fig. 5.2.4 Enthalpy–entropy diagram for air

In an attempt to give a physical meaning to the concept of entropy we may note the analogy between the two "non-properties", heat and work:

$$dQ = T \, ds$$
$$dW = p \, dV$$
$$= F \, dl$$

Each may be represented as the product of an intensive and an extensive property (work may also be represented as the product of force and distance, neither of which are thermodynamic properties in the accepted sense, although clearly closely related to pressure and volume).

We also note that in the same way that work is represented by areas on the p-V diagram, heat is represented by areas on the T-s diagram. One of the accepted verbal definitions of entropy is:

"The co-ordinate with temperature of heat".

Fig. 5.2.5 shows representations of various important thermodynamic processes on the T-s diagram for a gas.

Fig. 5.2.5(a) shows the isothermal transfer of heat to a gas:

$$T(s_2 - s_1) = Q_{12} + W_l$$

If no irreversibility is present, $W_l = 0$, and the rectangular area cross-hatched represents the heat supplied in the course of the process (for this to be the case the temperature scale must start at absolute zero).

Fig. 5.2.5(b) represents an isentropic expansion, one in the course of which no heat

101

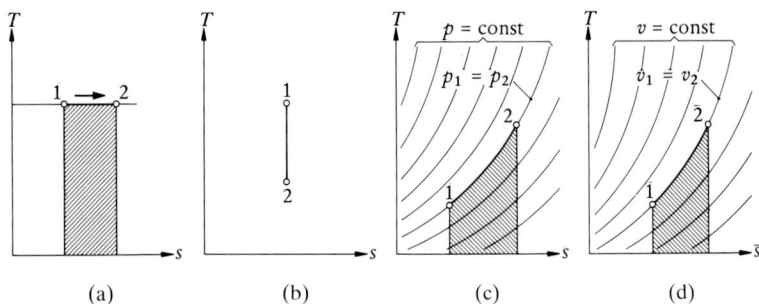

Fig. 5.2.5 Representation of various thermodynamic processes applied to a gas on the *T-s* diagram:
- (a) Isothermal heat transfer
- (b) Isentropic expansion
- (c) Addition of heat at constant pressure
- (d) Addition of heat at constant volume

enters or leaves the gas, and no irreversibility (i.e. internal friction loss) occurs. This process is represented by:

$$s_2 = s_1$$

Many real processes—for example, expansion in turbines and compression and expansion in i.c. engines—approximate closely to isentropic processes.

Fig. 5.2.5(c) represents an isobar, the supply of heat to a gas at constant pressure. *T-s* diagrams show constant pressure parameter lines, and the process is indicated as following such a line.

For permanent gases (gases in a state remote from the liquefaction point) we may write:

$$dQ = mc_p \, dT \tag{5.4}$$

where c_p = specific heat at constant pressure. Observation shows that over considerable ranges of pressure and temperature c_p may be regarded as constant, and we may calculate entropy changes as follows:

$$ds = \frac{dQ}{T} = mc_p \frac{dT}{T} \tag{5.5}$$

$$s_2 - s_1 = mc_p \log_e \frac{T_2}{T_1}$$

The parameter lines thus approximate to exponential curves. Once again the area beneath the curve on the *T-s* diagram represents the heat supplied to the gas during the process.

Fig. 5.2.5(d) represents the supply of heat to a gas at constant volume:

$$dQ = mc_v \, dT$$

where c_v = specific heat at constant volume. It is always the case that $c_p > c_v$, since when heat is supplied at constant pressure the heat equivalent to the displacement work performed during the expansion of the gas must also be provided. The equation for the entropy change is analogous to (5.5):

102

$$s_2 - s_1 = mc_v \log_e \frac{T_2}{T_1} \tag{5.6}$$

For solids and liquids for which the thermal expansion relative to that of gases is negligible we may write $c_p \sim c_v = c$:

$$s_2 - s_1 = mc \log_e \frac{T_2}{T_1} \tag{5.7}$$

5.3   The Carnot Cycle

In 1824, S. Carnot proposed an ideal cycle which, if it could be realized, would yield an efficiency of conversion of heat into work equal to that predicted by the Second Law:

$$\eta = \frac{T_1 - T_2}{T_1}$$

It was a measure of Carnot's genius that he deduced this efficiency at a time when the First Law had in fact not yet been formulated.

Carnot visualized the use of an ideal gas (see p. 166) as the working fluid; the argument is slightly easier to follow and yields the same result if we substitute water and steam for the ideal gas.

The successive phases of the cycle are illustrated in Fig. 5.3.1, which also shows the accompanying p-V and T-s diagrams; essentially the cycle is built up of two isothermal and two isentropic processes, see Fig. 5.2.5(a) and (b). The machine in which the process is carried out is envisaged as consisting of a cylinder and piston of perfectly insulating material, the end of the cylinder consisting of a thin, perfectly conducting diaphragm. It is assumed that leakage and friction are absent and that the process takes place slowly so that losses associated with fluid friction are also eliminated. We assume, in fact, that all the sources of irreversibility listed in Table 5.1.1 are absent.

The machine is associated with two reservoirs of heat: a "source" at temperature T_1 and a "sink" at a lower temperature T_2.

The cycle may be described in terms of the four following phases.

Phase 1: Fig. 5.3.1(a)

The piston is at the bottom of the cylinder, the whole of the clearance space being filled with water at temperature T_1. The source of heat at temperature T_1 is placed in contact with the end of the cylinder, and heat flows without temperature difference into the water, which evaporates at constant pressure p_1, causing the piston to rise and perform work against a resisting force $F_1 = p_1 A$, where $A =$ piston area. This process is represented on the p-V and T-s diagrams in the figure.

Phase 2: Fig. 5.3.1(b)

When all the water in the cylinder has evaporated, giving saturated steam, the heat source is removed and an insulating cover applied to the end of the cylinder. The piston continues to move outwards against a steadily falling resisting force, until the

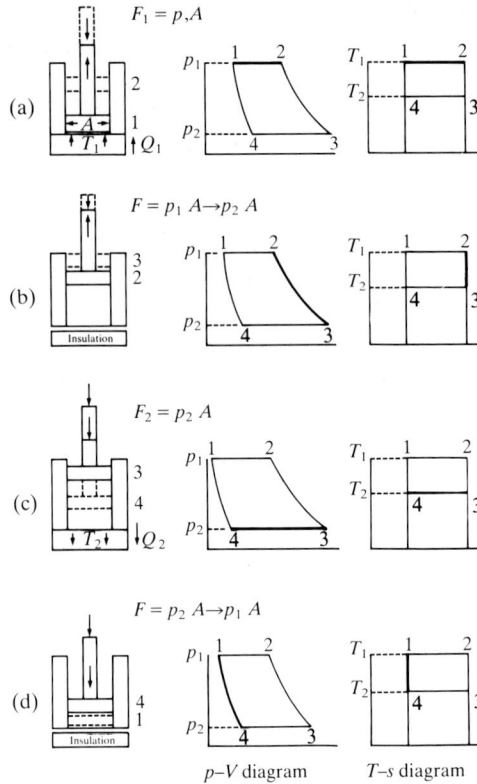

Fig. 5.3.1 The Carnot cycle for steam

pressure of the steam has fallen from p_1 to p_2 and the temperature from T_1 to T_2. No heat transfer to or from the steam takes place; the process is adiabatic as well as isentropic.

Phase 3: Fig. 5.3.1(c)

When the pressure of the steam has fallen to p_2 the insulation is removed and the heat sink at temperature T_2 is placed in contact with the cylinder end. The piston now moves downwards, driven by a constant force $F_2 = p_2 A$, and heat Q_2 is transferred without temperature difference to the sink, while the steam progressively condenses at constant pressure p_2.

Phase 4: Fig. 5.3.1(d)

At the appropriate point in the cycle the sink is removed and the insulating cover again applied to the cylinder end. The piston is driven inwards under an increasing force, resulting in a compression of the steam which is again adiabatic and isentropic. Finally all the steam is recondensed at the higher pressure p_1 and we arrive back at the conditions obtaining at the start of the cycle.

As, by definition, no losses have taken place, we may write in accordance with the First Law:

$$Q_1 - Q_2 = W \tag{5.8}$$

The net heat withdrawn from the reservoirs must have appeared as work.

We apply the Second Law in the form of equation (5.3), noting that $W_1 = 0$:

$$s_2 - s_1 = \frac{Q_1}{T_1} = \frac{Q_2}{T_2}$$

or

$$Q_2 = Q_1 \frac{T_2}{T_1}$$

$$W = Q_1 \left(1 - \frac{T_2}{T_1}\right) = Q_1 \left(\frac{T_1 - T_2}{T_1}\right) = \eta_c Q_1 \tag{5.9}$$

This equation is, in fact, Formulation G of the Second Law given in Table 2.6.1. The maximum possible efficiency of a machine operating between temperatures T_1 and T_2 is the Carnot efficiency.

Fig. 5.3.2(a) shows the Carnot cycle on the T-s diagram, while Fig. 5.3.2(b) shows a cyclic process of generalized form. This does not, as is the case with the Carnot cycle, operate between two isotherms and two isentropics, and its efficiency is necessarily less than that of a Carnot cycle operating between the same temperature limits. The work performed by the machine is represented by the area $W = Q_1 - Q_2$ and the efficiency $\eta = W/Q_1$ is evidently less than that corresponding to the Carnot process.

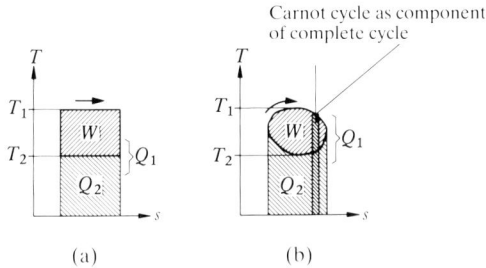

Fig. 5.3.2   Representation of thermodynamic cycles on the T-s diagram:
(a) Carnot cycle
(b) Generalized cycle

The same conclusion may be reached by visualizing the cycle of Fig. 5.3.2(b) as being built up from a series of Carnot cycles, some of which will necessarily operate between narrower temperature limits than T_1 and T_2 and will thus have lower efficiency.

5.4   Availability

It will be apparent that it is possible to classify different kinds of energy in accordance with the degree to which they are available for conversion into other forms. These fall into three categories:

(a) Forms of energy such as chemical, electrical and mechanical that are at least theoretically mutually convertible without loss. The losses associated with irreversibility that arise in the course of such transformations may be reduced to a minimum by appropriate technical means and are in principle avoidable.

(b) Heat and internal energy, the availability of which for conversion into other forms is dependent on temperature and governed by the requirements of the Second Law.

(c) Heat that is intrinsically unavailable for conversion into other forms. This is represented principally by the internal energy of the surroundings, for which any degree of availability would be dependent on the existence of a sink at lower temperature than the surroundings.

The simplest case concerns the availability of a heat reservoir at temperature T_1. It is important to be clear as to what constitutes a heat reservoir. Examples are:

(a) A nuclear reactor, the temperature T being the maximum working temperature of the heat transfer fluid.

(b) A mass of molten metal which releases heat at constant temperature in the process of freezing.

The definition of a heat reservoir is as follows: a body of infinite dimensions and perfect conductivity which is capable of releasing any required quantity of heat with an infinitesimal accompanying fall in temperature. Such a theoretical heat reservoir cannot be represented by, for example, 1000 kg of iron at a temperature of 500°C. Such a mass contains a great deal of heat (~ 250 MJ relative to atmospheric conditions), but the process of withdrawing heat results in a fall in temperature and the mass is then no longer at the same temperature.

The measure of the availability of a quantity of heat Q in a reservoir at temperature T_1 that meets the above conditions is given simply by the Carnot cycle efficiency:

$$\text{Available energy} = Q\left(\frac{T_1 - T_0}{T_1}\right)$$

where $T_0 =$ lowest available temperature of heat sink.

Availability is thus a function not only of the source but of its surroundings. In most instances $T_0 =$ atmospheric temperature, but special cases can be distinguished. For example, a few successful attempts have been made to exploit the (very limited) temperature differences that exist between the water near the surface of the ocean and at great depths by devising machines that operate on a vapour cycle (see Section 6.5) between these temperature limits. $T_1 - T_0 \sim 20°C$, implying a Carnot cycle efficiency $\eta_C \sim 6$ percent, and a realizable efficiency of much less than this, but the heat reservoir (and the heat sink) are both of virtually infinite extent and the heat supply is "free".

An important example concerns the availability of unit mass of compressed fluid at conditions (p_1, T_1). Imagine the fluid to be contained in a cylinder closed by a piston. It is required to calculate the work that can ideally be performed by the fluid in the course of a process terminating at the conditions of the surroundings (p_0, T_0).

Applying the First Law, equation (2.5):

$$u_1 - u_0 = Q_{10} + W \tag{5.10}$$

where u_1, u_0 = initial and final internal energies respectively, Q_{10} = heat transferred from the fluid to the surroundings, and W = work performed by the fluid.

In defining the availability, we take into account the fact that the useful work performed by the fluid in expanding from an initial volume v to final volume v_0 is reduced by the work performed by the piston upon the surroundings. The net available work then becomes:

$$W_A = W - p_0(v_0 - v_1) \qquad (5.11)$$

Combining equations (5.10) and (5.11),

$$W_A = u_1 - u_0 - Q_{10} - p_0(v_0 - v_1) \qquad (5.12)$$

In order to determine W_A we need to establish the amount of heat Q_{10} transferred from the fluid to its surroundings. By definition, we assume that no irreversibility is involved; we have therefore to postulate a reversible process by which the fluid moves from state (p_1, T_1) to state (p_0, T_0). If we can determine one such process we know from the Second Law (Table 2.6.1, statements C and G) that any reversible process will have the same efficiency.

Fig. 5.4.1 shows one possible reversible process. The fluid is expanded isentropically until its temperature has fallen to T_0. The pressure will not normally equal p_0, but may be envisaged as being restored to this value by an isothermal compression in the course of which Q_{10} is transferred reversibly to the surroundings:

$$Q_{10} = T_0(s_1 - s_0)$$

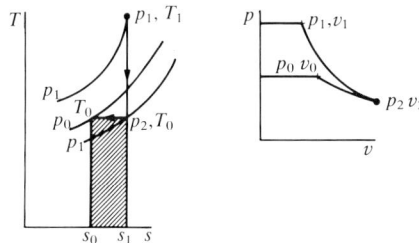

Fig. 5.4.1   Hypothetical reversible process for calculation of availability

where s_1, s_0 = initial and final entropy of the fluid. (Note that, even though the entropy of the fluid is shown as decreasing, the total entropy of fluid plus surroundings remains constant).

Substituting this value in equation (5.12):

$$W_A = u_1 - u_0 - T_0(s_1 - s_0) - p_0(v_0 - v_1)$$
or $$W_A = (u_1 + p_0 v_1 - T_0 s_1) - (u_0 + p_0 v_0 - T_0 s_0) \qquad (5.13)$$
$$= A_1 - A_0$$

where $A = (u + p_0 v - T_0 s)$ is known as the non-flow availability function.

As a simple example, consider the availability of a source of heat represented by 100 kg of water at a temperature of 60°C and a pressure of 100 000 N/m², in surroundings in which the pressure is the same and the temperature 20°C.

Using data from Steam Tables, we may write:

$$u_1 = 251 \cdot 2 \text{ kJ/kg}$$
$$s_1 = 0 \cdot 8309 \text{ kJ/kg.K}$$
$$u_0 = 84 \cdot 0 \text{ kJ/kg}$$
$$s_0 = 0 \cdot 2963 \text{ kJ/kg.K}$$
$$p_0 v_0 = p_1 v_1, \quad \text{very nearly}$$

Then, from equation (5.13),

$$W_A = [(251 \cdot 2 - 293 \times 0 \cdot 8309) - (84 \cdot 0 - 293 \times 0 \cdot 2963)]$$
$$= 1056 \cdot 2 \text{ kJ}.$$

Conversely, a minimum expenditure of 1056·2 kJ of mechanical work would be required to heat 100 kg of water from 20°C to 60°C. This would require a heat pump (see Experiment 6) that would operate without losses.

For a steady-flow open system the corresponding available work is given by:

$$W_B = (h_1 - T_0 s_1) - (h_0 - T_0 s_0) \tag{5.14}$$
$$= B_1 - B_0$$

where $B = (h - T_0 s) = (u + pv - T_0 s)$ is known as the steady-flow availability function.

The analysis of processes and in particular of processes such as combustion, on the basis of availability (sometimes known as Second Law analysis) can yield important information. As an example, Fig. 5.4.2 represents the performance of a steam power

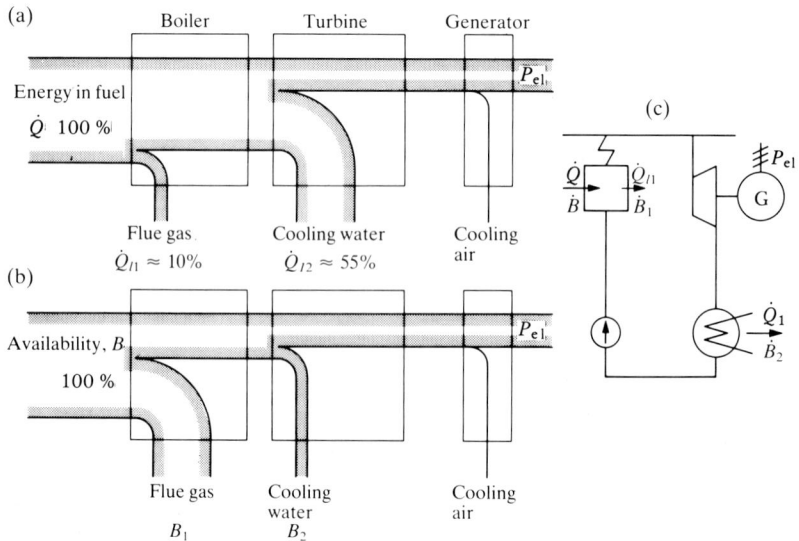

Fig. 5.4.2   Energy and availability for a steam power plant
(a) Energy balance
(b) Availability analysis

plant, both as an energy balance and as an availability analysis. The striking difference between these two analyses is that, while the energy balance shows that 55 percent of the energy released by combustion in the boiler finds its way into the condenser cooling water and only 10 percent into the flue gas, in the case of the availability balance the figures are reversed.

The available work in the exhaust steam is small because its temperature only exceeds that of the surroundings, in this case represented by the condenser cooling water supply, by the amount necessary for heat transfer.

In the boiler, on the other hand, the maximum temperature of the steam which determines the efficiency of the power-generating cycle is very much less than the temperature generated by the combustion of the fuel. The efficiency of the process may be improved by increasing the maximum steam temperature.

Note that the energy in the fuel is in principle completely available; reaction in a fuel cell, if it were to become practicable with readily available fuels, would give rise to much higher efficiencies of conversion into electrical power (100 percent efficiency is theoretically possible) than can be achieved by burning the fuel in a boiler, even with the most extreme steam conditions that are at present practicable.

5.5   Experiment 6: the Heat Pump

One statement of the Second Law of Thermodynamics, F, Table 2.6.1, implies that it is possible to transfer heat from a reservoir at a lower temperature T_2 to a higher temperature T_1 by the expenditure of work. The efficiency of this process is the reverse of the Carnot efficiency:

$$\dot{Q}_2 = \frac{\dot{W}T_2}{T_1 - T_2}; \quad \frac{\dot{Q}_1}{\dot{Q}_2} = \frac{T_1}{T_2} \tag{5.15}$$

The form of this equation implies that the smaller the temperature difference the greater the amount of heat that may be transferred for a given expenditure of work.

As early as 1850, Lord Kelvin proposed a reversed heat engine of this type, which he called a heat pump, for the warming or, by implication, the cooling of buildings. The difference between the desired temperature in the building and that of the outside air is in most cases fairly small, and it should thus be possible to transfer a considerable quantity of heat for the expenditure of comparatively little work. In the event the heat pump has still found only restricted application; the possible energy savings have proved insufficiently attractive when compared with the cost and complexity of the equipment required.

This situation is changing rapidly with the increasing cost and scarcity of fuel and "total energy systems" comprising an engine-driven heat pump in which heat is also transferred to the space to be warmed from the engine cooling water and exhaust gas, may be expected to find wide use in the future. Such systems may have an "efficiency", as measured by the ratio of the energy transferred to the building to be heated divided by the energy content of the fuel, of much more than 100 percent.

The increasing importance of nuclear power will also affect the situation. It is inherently inefficient to use electricity, generated with a thermal efficiency unlikely to exceed 40 percent, for direct heating purposes; it is more efficient to burn the fuel used in the power station on site to heat the building directly. This option is not open with nuclear power.

Refrigerators operate on the same principle as the heat pump and have, of course, found extremely wide application.

The heat pump is one of the few practical machines the performance of which may be related directly to the Carnot cycle, and the authors made an analysis of the performance of such a unit. They used a commercially available air conditioner which has the special feature that the source of heat is provided by a flow of water; for most small machines of this type heat is drawn from the external atmosphere. The use of water makes it a simpler matter to draw up an accurate heat balance. The internal connections in the authors' unit could also be rearranged to enable it to act as an air conditioner, transferring heat from the room in which it is installed to the circulating water.

Fig. 5.5.1 shows a general view of the apparatus and Fig. 5.5.2 shows the circuit and instrumentation. The apparatus is built round a hermetically sealed refrigeration system driven by an electric motor and employing the refrigerant R22 (see p. 132). The refrigerant is compressed, condensed at a relatively high temperature with the release of heat, expanded through a throttling valve, evaporated at a lower temperature and pressure with the absorption of heat, and again re-compressed.

Such refrigeration processes, in principle identical with the heat pump process, are described in Chapter 6. At present we are interested in the relation between the overall process and the corresponding ideal reversed Carnot cycle.

Referring to equation (5.15), the input of work $\dot{W}$ is represented by the electrical power required to drive the refrigerator compressor motor, together with the small amount of power required to drive a fan which circulates air through the apparatus. The source of heat $\dot{Q}_2$ at the lower temperature T_2 is represented by the flow of circulating water through the heat exchanger in which the refrigerant is evaporated. The heat $\dot{Q}_1$ is transferred to the room in which the apparatus is installed by way of a further heat exchanger in which the refrigerant is condensed.

The air leaving the apparatus is discharged at comparatively high velocity through a circular duct in which the velocity of flow is measured by a pitot tube with static tapping permitting measurement of the mass flow rate. The mean air velocity bears a known relationship to that indicated by the pitot tube ([11]).

The air and cooling-water inlet and outlet temperatures are measured by mercury-in-glass thermometers, while the temperatures of the refrigerant at inlet and outlet to the compressor and at inlet and outlet to the evaporator and condenser are measured by thermocouples. Cooling-water flow is measured by a Rotameter, and an auxiliary electrical heater is provided to maintain the circulating water inlet temperature at a sufficiently high level; if this is less than 10°C there is the possibility of ice formation in the heat exchanger.

Fig. 5.5.1 Experimental heat pump and air cooler

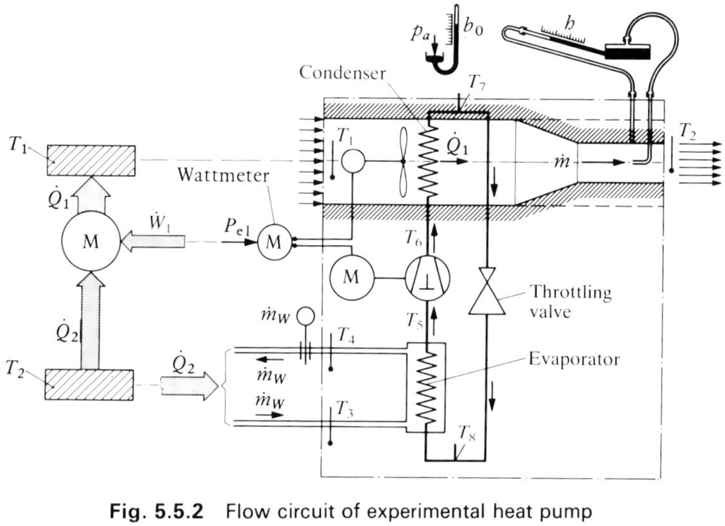

Fig. 5.5.2 Flow circuit of experimental heat pump

111

5.5.1 Measurements and Calculations

Table 5.5.1 shows a typical set of observations which are analysed below with reference to the control volume and energy flow diagram of Fig. 5.5.3.

Table 5.5.1 Experiment 6: the Heat Pump: Observations

t_1/T_1	21·1°C	294·1K
t_2/T_2	42·4	315·4
t_3/T_3	11·7	284·7
t_4/T_4	5·6	278·5
t_5/T_5	3	276·0
t_6/T_6	53·0	326·0
h mmH$_2$O	43·7	
$\dot{m}_w$ kg/h/kg/s	195	0·0542
E W	1280	
E_F W	300	

Ambient temperature   21 1°C
Barometer   750 mmHg

The air mass flow rate in the authors' apparatus is given by:

$$\dot{m} = 2\cdot371 \, kd^2 \sqrt{\frac{hh_0}{T_2}}$$

where $k = \dfrac{\text{mean velocity in duct}}{\text{velocity indicated by pitot tube}} = 0\cdot96$

$d =$ duct diameter
$h =$ dynamic head, mmH$_2$O
$h_0 =$ barometer, mmHg
$T_2 =$ air temperature

$$\dot{m} = 2\cdot371 \times 0\cdot96 \times 0\cdot073^2 \sqrt{\frac{43\cdot7 . 750}{315\cdot4}}$$

$$= 0\cdot1236 \, kg/s$$

Enthalpy of air entering heat pump:

$$\dot{Q}_1 = \dot{m}c_p(T_1 - 273) = 0\cdot1236 . 1005 . 21\cdot1$$
$$= 2621 W$$

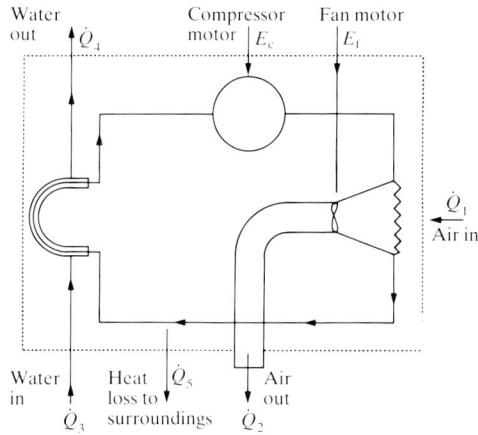

Energy Flow Diagram

Fig. 5.5.3 Energy flow diagram of heat pump

Enthalpy of air leaving heat pump:

$$\dot{Q}_2 = \dot{m}c_p(T_2 - 273) = 0{\cdot}1236 \,.\, 1005 \,.\, 42{\cdot}4$$
$$= 5267\text{W}$$

Enthalpy of water at inlet:

$$\dot{Q}_3 = \dot{m}_w c_w (T_3 - 273)$$

where c_w = specific heat of water = 4187 J/kgK
$\dot{Q}_3 = 0{\cdot}0542 \,.\, 4187 \,.\, 11{\cdot}7$
$= 2655$ W

Enthalpy of water at exit:

$$\dot{Q}_4 = \dot{m}_w c_w (T_4 - 273)$$
$$= 0{\cdot}0542 \,.\, 4187 \,.\, 5{\cdot}6$$
$$= 1271 \text{ W}$$

Power input to compressor motor:

$$E_C = 980 \text{ W}$$

Power input to fan:

$$(E_F) = 300 \text{ W}$$

The First Law (s.f.e.e.) may be written:

$$(\dot{Q}_2 - \dot{Q}_1) + (\dot{Q}_4 - \dot{Q}_3) + \dot{Q}_5 = E_C + E_F$$
$$2646 - 1384 + 18 = 980 + 300$$

The stray heat losses from the apparatus, $\dot{Q}_5$, are calculated by difference and are evidently very small.

113

The analysis may also be presented in the form of an energy balance:

Electrical energy, compressor motor	980 W	Heat to air	2646 W
Electrical energy, fan motor	300 W	Heat to surroundings	18 W
Heat from circulating water	1384 W	(by difference)	
	2664 W		2664 W

5.5.2 Discussion of Results

The analysis shows that for the expenditure of 1280 W of electrical power the heat pump has transferred 2646 W to the air circulated through it, the difference having been drawn from the circulating water. The heat pump is thus a considerably more efficient source of warmth than an electrical resistance heater; it must, however, be recognized that the electrical power expended has itself probably been produced by the combustion of fuel in a power station with an overall efficiency in the region of 30 percent. Combustion of the same fuel in a local heating unit could be a more efficient process.

The efficiency of a heat pump is described by a coefficient of performance, representing the ratio between the useful heat delivered to the surroundings and the power expended:

$$C.P. = \frac{\dot{Q}_2 - \dot{Q}_1}{E_C + E_F}$$

For the present test:

$$C.P. = \frac{2646}{1280} = 2 \cdot 07$$

A better measure of the performance of the heat pump is a comparison between the coefficient of performance and the coefficient of performance of an ideal reversed Carnot cycle machine operating between the same temperature limits. These may be taken as:

Mean air temperature   304·8K

Mean water temperature   281·7K

$$C.P. \text{ (Carnot)} = \frac{304 \cdot 8}{304 \cdot 8 - 281 \cdot 7} = 13 \cdot 2$$

Power input, ideal Carnot 2646 ÷ 13·2 = 200 W

This leads to the somewhat disappointing conclusion that the efficiency of our heat pump is only about 16 percent of that of an ideal reversed Carnot cycle machine operating between the same temperature limits.

This is a typical situation in which an availability or Second Law analysis can yield valuable information; it should be possible to determine the reasons for the very large difference between the ideal and actual performances.

It will be apparent that a large part of this difference must arise from the fact that the refrigerator operates between a greater range of temperatures than that represented by the difference between those of the circulating water and the warmed air. The coefficient of performance of a reversed Carnot cycle machine operating between the temperatures T_6 of the refrigerant in the condenser and T_5 in the evaporator is given by:

$$\text{C.P. (Carnot refrigerator)} = \frac{T_6}{T_6 - T_5} = \frac{326}{326 - 276} = 6.52$$

Power input: $2646 \div 6.52 = 406$ W

Clearly a substantial part of the deficiency in the performance of the machine is associated with the temperature differences necessary to transfer heat from the circulating water to the refrigerant and from the refrigerant to the warmed air.

If we further subtract the electrical power absorbed by the fan from the power consumption of the machine (noting that this power appears as part of the heat imparted to the air), we may derive a further value for what may be described as the "internal" coefficient of performance of the machine:

$$\text{C.P. (internal)} = \frac{2646 - 300}{1280 - 300} = 2.39$$

We may conclude that, when allowance is made for the necessary temperature differences to effect heat transfer and for the power absorbed in the fan, the efficiency of our heat pump is some 37 percent of that of a reversed Carnot cycle machine.

It is a reasonable assumption that the overall efficiency of the refrigerator compressor and motor, defined as the ratio of the indicated work performed by the compressor on the refrigerant (p. 174) to the electrical power input to the motor, is not more than 50 percent. This leads to a relative efficiency when compared with a reversed Carnot machine in the region of 74 percent, indicating that we have identified the principal sources of inefficiency in our heat pump.

5.5.3 Further Experiments and Questions

(a) Examination of the Performance of the Heat Pump over a Range of Temperatures

By varying the circulating water inlet temperature and also by allowing the temperature of the laboratory to rise to high values it is possible to investigate the performance of the apparatus over a range of temperatures. This is of particular interest since the results provide some direct evidence of the validity of the Carnot cycle as a measure of the efficiency of real thermodynamic processes. Fig. 5.5.4 shows the result of such a study in the form of a plot of the inverse of the coefficient of performance against the temperature range. The various curves represent the following:

(1) Performance of ideal reversed Carnot cycle machine operating between the temperature limits T_1 and T_2.

(2) Ideal reversed Carnot cycle machine operating between refrigerant temperature limits (allowance for heat transfer temperature difference of 25°C).

(3) Performance of actual machine.

(4) Performance of actual machine after allowance for fan power.

(5) Performance of actual machine assuming compressor efficiency equals 50 percent.

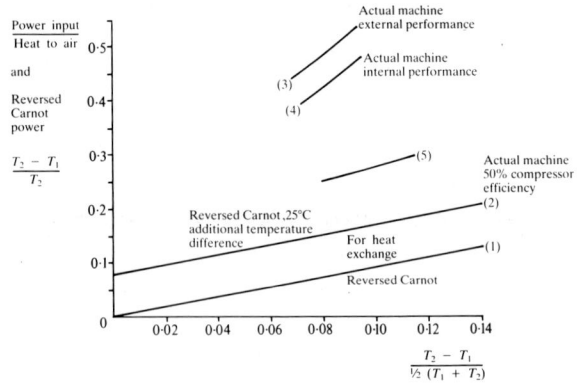

Fig. 5.5.4   Effect of temperature range on heat pump performance

It will be apparent that curves (1) and (5) lie substantially parallel.

(b) Operation of the Heat Pump in the Reversed Mode as a Cooler, transferring Heat from the Air to the Circulating Water

It is to be noted that in this case it is necessary to allow for the latent heat of condensation of any moisture deposited from the air in its passage through the machine (see Chapter 7).

6

Properties of Working Fluids: Thermodynamic Cycles

6.1 Properties of Fluids

We have already seen that thermodynamic cycles and processes do not take place so to speak *in vacuo*; a working fluid is involved. This working fluid may be a liquid, a vapour, a gas, a mixture of these phases or, in rare instances, a solid. Comprehensive data on the thermodynamic properties of the fluids commonly employed in thermodynamic processes is thus essential to enable us to design and analyse real processes. Engineering thermodynamics is principally concerned with the following properties, here described by the symbols with which the reader will already be familiar: p, v, T, u, h, s. Certain other thermodynamic properties have been defined; their principal significance is in connection with chemical reactions.

The first three properties, pressure, specific volume and temperature, may be observed without difficulty; the next, the specific internal energy, may also be measured, though the process is more complicated. The final properties, specific enthalpy and specific entropy, may be calculated on the basis of data for the first four properties. (The word specific, implying that the value quoted refers to unit mass of the fluid, is usually omitted.)

Thermodynamic properties must be determined empirically; they cannot be deduced on the basis of the laws of thermodynamics and, except in the special case of a perfect gas, the relation between them cannot be expressed in the form of a simple algebraic function.

The Two Property Rule, the proof of which will not be given here, states that if any two independent properties of a pure fluid are known, the remaining properties may be determined. This rule provides the basis for the presentation of thermodynamic data in the form of charts showing the relationship between two properties and carrying parameter lines corresponding to specific values of the other properties.

The early development of thermodynamics was linked closely to the development of the steam engine, and the thermodynamic properties of water/steam were investigated at a relatively early date. We may imagine a process for determining the properties of this substance, Fig. 6.1.1(a). One kilogram of water is placed in a cylinder closed by a piston and the temperature is lowered until the water freezes. We now transfer measured quantities of heat to the ice, perhaps by an electrical resistance heater, and maintaining the pressure on the piston constant, we observe the relations between the temperature and specific volume of the fluid and the heat supplied. Under

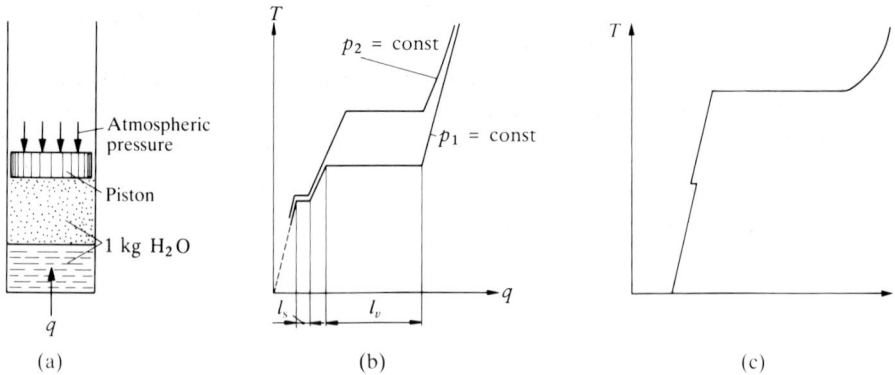

Fig. 6.1.1 Boiling of water at constant pressure:
(a) Hypothetical apparatus
(b) Relation between heat supplied and temperature
(c) Volume changes with temperature

the conditions we have specified, the heat supplied $q = h$, the enthalpy (in charts and tables the datum point for enthalpy is usually taken as $0°C$).

Fig. 6.1.1(b) shows the result of the experiment we are considering in the form of a T-q diagram. Initially the addition of heat causes the temperature of the ice to rise until a temperature of $0°C$ reached. At this point the addition of further heat gives rise to no increase in temperature but the ice begins to melt. The amount of heat required to bring about the change of phase from ice to water is known as the latent heat of fusion. Further heating results in a rise of temperature until, at a temperature of $100°C$, the water begins to boil. At this point a second change of phase occurs, from water to steam, accompanied by a large increase in volume and the absorption of a large quantity of heat at constant temperature, the latent heat of vaporization, l_v. Finally, the whole of the water is vaporized and the cylinder is full of saturated steam. Further increase in the heat supplied results in a rise in temperature and the steam becomes superheated.

Fig. 6.1.1(c) shows the corresponding changes in volume. The ice expands slightly as its temperature is raised to $0°C$, but the melting of the ice is accompanied by a contraction. A further very slight contraction of the water occurs as its temperature rises to $4°C$, after which it expands slightly as the temperature rises to the boiling point. The behaviour of ice/water in this respect is anomalous; normal substances expand on melting. The increase in volume accompanying the change of phase from liquid to vapour is very much greater than is suggested in Fig. 6.1.1(c); at atmospheric pressure the specific volume of saturated steam is some 1600 times that of water. With further increase in temperature the volume of the superheated steam also increases.

If we progressively extract heat from the superheated steam the process follows precisely the same course without hysteresis. The steam condenses with release of the latent heat of condensation which precisely equals the latent heat of vaporization and with subsequent cooling the water freezes with release of the latent heat of freezing, in turn equal to the latent heat of fusion.

By repeating this process at a number of different pressures we may in principle build up a complete diagram of the thermodynamic properties of water.

118

Fig. 6.1.2 shows one of the standard methods of presenting thermodynamic data: as a v-T diagram. Principal features are the saturated liquid line and the saturated vapour line, defining respectively the discontinuities in curves such as that shown in Fig. 6.1.1(c). At the critical point, which for water corresponds to a temperature of 374°C, liquid and vapour lines meet, and above this temperature the liquid and vapour phases become indistinguishable.

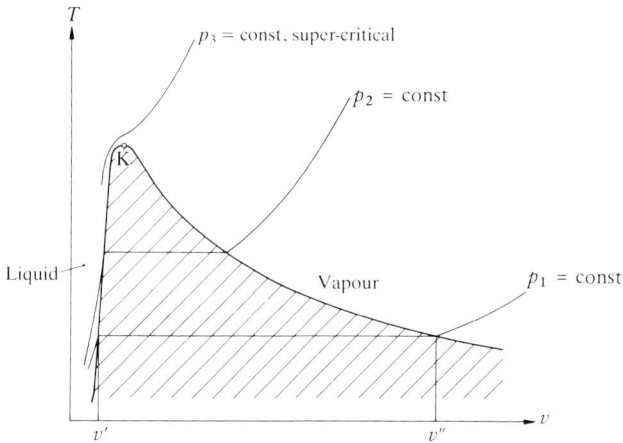

Fig. 6.1.2   Volume–temperature diagram for water

The latent heat of vaporization is zero at the critical point and at higher temperatures has no meaning. At temperatures above the critical it is no longer possible to liquefy any gas by compression alone; its temperature must also be reduced below the critical. While no precise distinction can be made between gases and vapours, the description gas is usually confined to fluids for which the critical temperature is very low (e.g. 154K for oxygen).

As the pressure is reduced, the temperature at which boiling occurs falls and the volume of vapour produced increases rapidly. In the case of water the melting point rises very slightly with reduction in pressure (for other fluids it falls) and eventually we reach a pressure and temperature at which the fusion and boiling temperatures are identical. At this triple point and at no other, ice, water and steam can exist together. The temperature of the triple point has been accepted internationally as a fixed point in the absolute temperature scale, see p. 26. The corresponding temperature and pressure are:

$$T = 273 \cdot 16 \text{K}, \ +0 \cdot 01°\text{C}$$
$$p = 611 \cdot 2 \text{ N/m}^2.$$

At pressures below that corresponding to the triple point ice sublimates directly into vapour without the intervention of a liquid phase.

Diagrams such as Fig. 6.1.2, showing the relationships between p, v and T, are not of great use in the representation and analysis of thermodynamic processes; information regarding the other three properties, u, h and s is required. This information is usually presented in the form either of the temperature–entropy diagram (a) or the enthalpy–entropy or Mollier diagram (b), Fig. 6.1.3. Fig. 6.1.4 shows the T-s diagram for steam in more detail, and Fig. 6.1.5 the Mollier diagram.

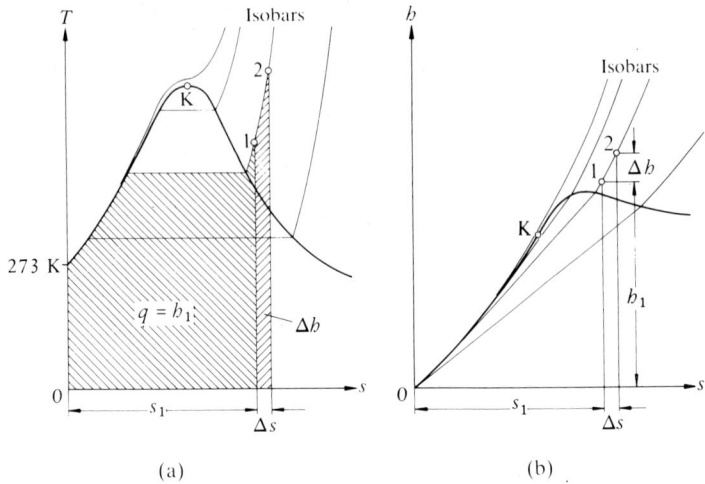

Fig. 6.1.3   Relation between temperature–entropy and enthalpy–entropy diagrams for steam. Areas under the isobars in the T-s diagram correspond to ordinates on the h-s diagram

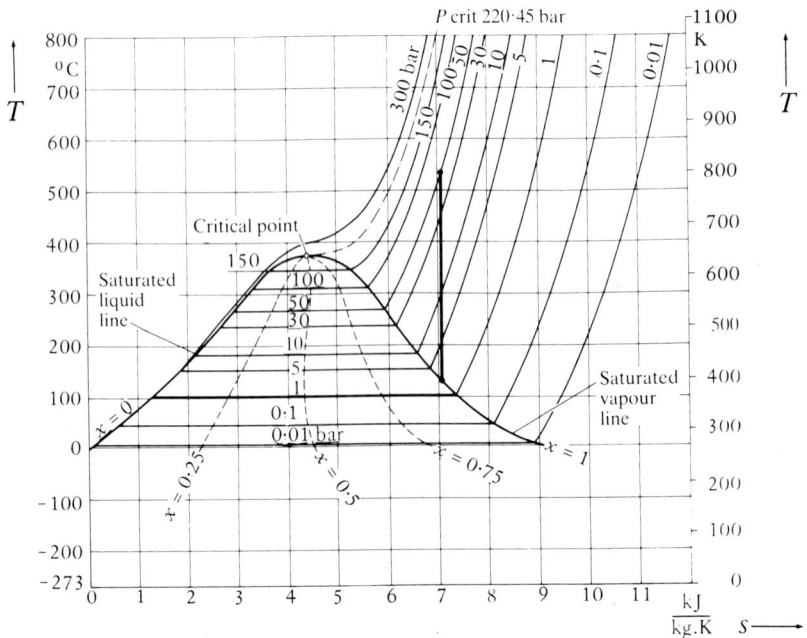

Fig. 6.1.4 T-s diagram for steam

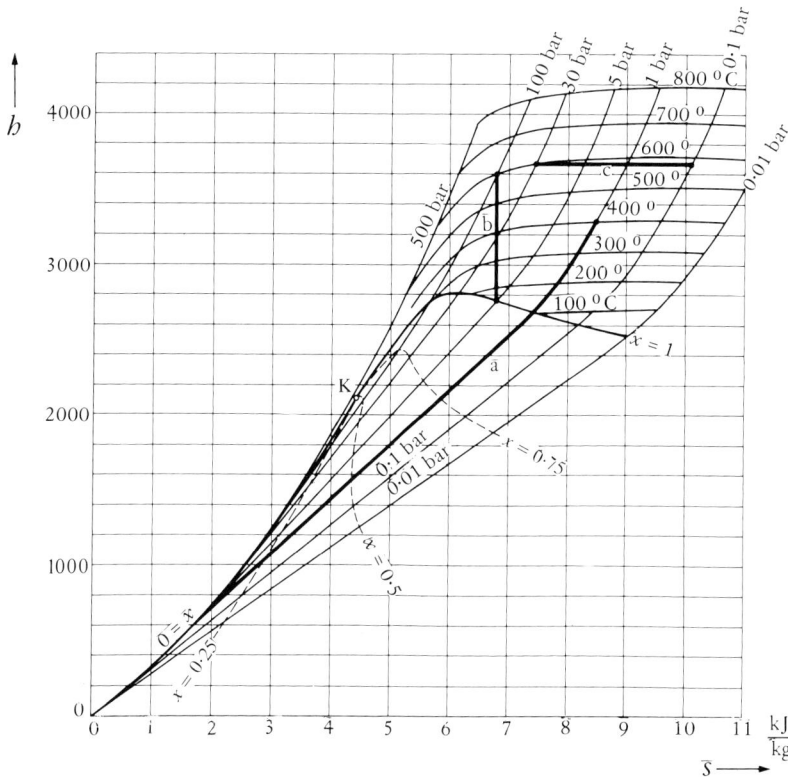

Fig. 6.1.5 Mollier diagram for steam

Practice regarding the choice of origin for the scales of s and h is not uniform. In the case of water the Steam Tables 1978 [18] specify that $s = 0$ and $h = 0$ for water at the triple point. It follows from the definition of enthalpy, equation (4.3), that at this point u has a very small negative value:

$$h = u + pV; \quad p = 611\cdot2 \, \text{N/m}^2, \quad V = 0\cdot001 \, \text{m}^3/\text{kg}$$
$$0 = u + 611\cdot2 \times 0\cdot001; \quad u = -0\cdot611 \, \text{Nm/kg}.$$

For some other fluids, for example dry air [19], the origin for both h and s is taken to be at the absolute zero of temperature. This involves difficult measurements in the region of the absolute zero and, since this temperature cannot actually be attained, certain theoretical assumptions. (The so-called Third Law of Thermodynamics states that for any pure substance the entropy approaches zero as the thermodynamic temperature approaches absolute zero.) The choice of origin is of little practical importance since we are generally only concerned with differences in the values of entropy and enthalpy.

The T-s diagram has two particular advantages: the reversible exchange of heat between fluid and its surroundings is represented by areas beneath the corresponding T-s curve while frictionless and adiabatic expansions or compressions are represented by vertical lines on the diagram.

121

An isentropic expansion or compression may also be represented by a vertical line on the Mollier diagram, with the further advantage that in this case the corresponding values of enthalpy may be read off directly and the adiabatic heat drop, $h_1 - h_2$ computed. The Mollier diagram is of particular application in the design of turbomachinery. A further advantage of the Mollier diagram is that heat exchanges at constant pressure, such as take place in a boiler, may be represented by successive points on an isobar. The application of the T-s and Mollier diagrams to a steam cycle is illustrated in the analysis of Experiment 7.

Mention should be made of a further method of presenting thermodynamic data, in this case as a p-h chart. This method of presentation is used for the presentation of refrigeration cycles and an example is given in Fig. 6.5.2, *below*.

As an alternative to the presentation of thermodynamic data in the form of diagrams, Tables of Properties have been prepared for fluids of thermodynamic importance, see, for example [19]. Properties are shown in the tables to a greater precision than is possible in a diagram, commonly to four significant figures, but the extraction of data from the tables involves interpolation and is laborious.

6.2 Steam Power Plant: the Rankine Cycle

In Section 5.3 we have described one possible realization of the Carnot cycle using steam as the working fluid. Our analysis showed all phases of the cycle as taking place in a single cylinder. This would evidently not be a practical solution and in reality each phase would be separate. Fig. 6.2.1 shows a possible arrangement in which the successive phases may be compared directly with Fig. 5.3.1, p. 104. The figure also shows directly the corresponding T-s diagram for the cycle.

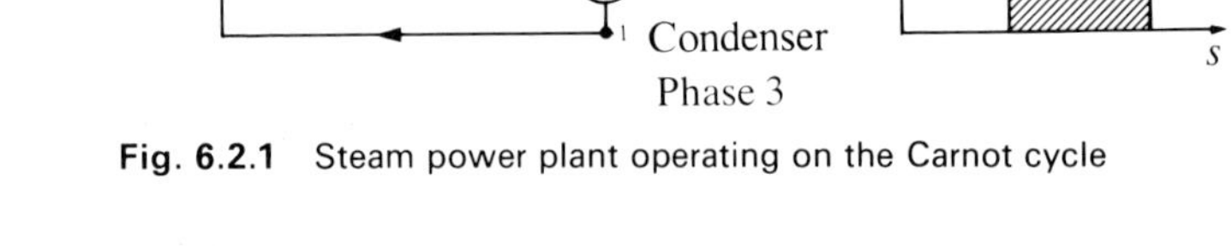

Fig. 6.2.1   Steam power plant operating on the Carnot cycle

The first phase, the transfer of heat at temperature T_1 and the generation of steam, takes place in the boiler. The second phase, isentropic expansion, takes place in an expander such as a steam turbine. The third phase, condensation of the steam with rejection of heat at temperature T_2, takes place in the condenser, and finally the steam is recompressed to boiler pressure by a compressor.

There are two principal reasons why the Carnot cycle is not used in practice. It would be very difficult to cut short the process of condensation at exactly the right point and then to compress efficiently a very wet mixture of condensate and steam.

122

Secondly, the power input required by the compressor would be a substantial fraction of the power output of the expander: the work ratio would be unfavourable.

$$\text{Work ratio} = \frac{\text{positive power output of expander}}{\text{negative power output of compressor}}$$

It will be evident that the lower the work ratio the more sensitive the efficiency of the whole cycle is to irreversibilities in the individual processes. In a cycle with low work ratio the pV diagram is "thin". A fatal objection to the use of a gas as the working fluid in any attempt to realize the Carnot cycle in practice is the extreme "thinness" of the pV or "indicator" diagram. The compression work is only slightly less than that realized in the expansion.

These objections to the Carnot cycle for steam are met by the adoption of the Rankine cycle, Fig. 6.2.2. In this cycle the condensation process is carried to completion and the water is then returned to the boiler by way of a feed pump, which absorbs very little power when compared with that required by the Carnot cycle compressor. This results in a substantial improvement in the work ratio, and for moderate boiler pressures the "feed pump" term corresponding to the work expended in forcing the feed water into the boiler is so small that it is usually neglected. The Rankine cycle has a slightly lower theoretical thermal efficiency than the corresponding Carnot cycle, but this is of little significance when weighed against the practical advantages of the cycle.

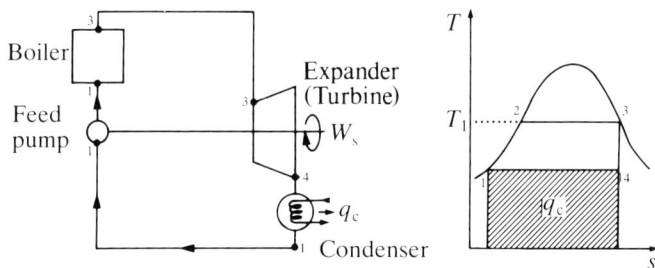

Fig. 6.2.2 Steam plant operating on the Rankine cycle

In actual steam power plants the efficiency of the cycle is invariably improved relative to that of the elementary Rankine cycle by the use of superheat, Fig. 6.2.3. The heating process in the boiler is not stopped at a point corresponding to saturated steam but is continued until the steam reaches a temperature limited only by metallurgical considerations regarding the material of the boiler and turbine blades. The effect of this modification is greatly to increase the available adiabatic heat drop and hence the potential power output obtainable from a given mass of steam at the expense of a comparatively modest increase in the heat consumption of the boiler. Substantial improvements in the ideal efficiency of the cycle may be achieved.

Heat transfer between states 1 and 3 takes place in the boiler. Further heat transfer between states 3 and 4 takes place in the superheater, an array of tubes external to the boiler proper. Despite precautions, the violent boiling process always results in the

123

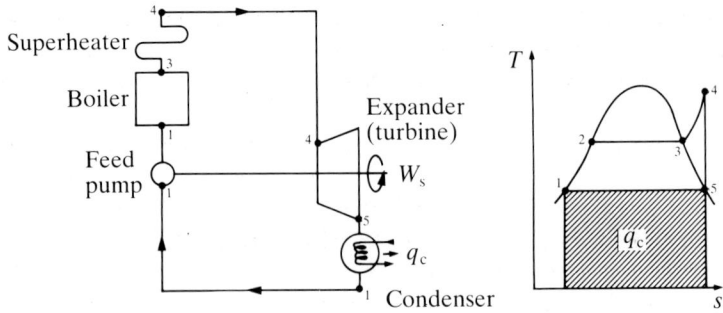

Fig. **6.2.3** Steam plant operating on Rankine cycle with superheat

carry-over of some droplets of water with the steam from the boiler to the superheater or, in the absence of a superheater, direct to the expander. The consequence is that state 3 does not in practice precisely correspond to dry saturated conditions.

After superheating, the steam expands through the turbine, performing work, entering the turbine in state 4 and leaving in state 5. The expansion is assumed to be adiabatic and isentropic. The work performed is equal to the change in enthalpy of the steam. It will be observed that in Fig. 6.2.3 the steam conditions at state 4 have been so chosen that the steam leaving the turbine at state 5 is in the dry saturated condition. It then enters the condenser, where the whole of its latent heat is transferred to the cooling water with a return to state 1 which lies on the saturated liquid line. The thermal efficiency of the simple Rankine cycle is further improved in practice by the introduction of one or more reheat stages in the cycle, Fig. 6.2.4. The steam is withdrawn from the turbine at a pressure intermediate between the inlet and condenser pressures and returned to the furnace, where its temperature and hence degree of superheat is increased. The introduction of reheat stages improves both the theoretical efficiency of the cycle and the work ratio, while it has the further important advantage of maintaining the steam in a superheated condition through most of its passage through the turbine. In the absence of sufficient superheat and reheat, the steam during part of its expansion from state 4 to state 5 will be in the wet condition

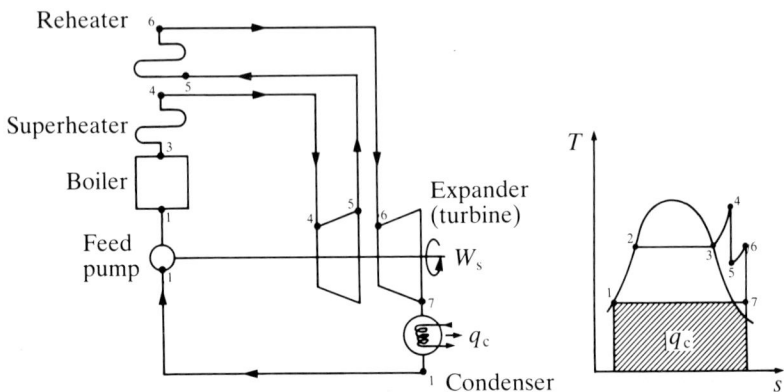

Fig. **6.2.4** Rankine cycle with superheat and reheat

(i.e. line 4–5, Fig. 6.2.3, will intersect the saturated vapour line on the T-s diagram). Wet steam may contain droplets of liquid water in suspension, and at the high velocities associated with flow in the steam turbine these give rise to severe blade erosion.

T-s diagrams for water show lines of constant dryness fraction (see Fig. 6.1.4), the significance of which will be self-evident. A dryness fraction of 75 percent implies that the fluid in this condition is a mixture of 75 percent saturated steam and 25 percent water at the same temperature. Saturation lines are plotted on the T-s diagram by drawing lines at constant temperature between the saturated liquid and the saturated vapour lines and dividing the interval in proportion to the dryness fraction. Points of equal dryness fraction at different temperatures are then joined to give a line of constant dryness fraction:

$$\text{Dryness fraction} = \frac{\text{mass of steam}}{\text{mass of steam} + \text{water}}$$

Provided the steam is only slightly wet at exit from the boiler (dryness fraction $\nleq \sim 0.9$), its condition may readily be determined by making use of a throttling process. We have already seen (Section 4.1) that throttling is characterized by a constant value of enthalpy. If, therefore, we withdraw a sample of steam from the boiler and throttle it to a lower pressure we may represent the process by a line of constant enthalpy on the T-s diagram or by a horizontal line on the Mollier diagram, Fig. 6.2.5. If the steam is initially sufficiently dry after throttling, it will have entered the superheat region, and an observation of the temperature will enable the degree of superheat to be determined and a line representing the throttling process plotted on the T-s or h-s diagram. The dryness fraction of the steam may then be read off the diagram; alternatively data from the steam tables may be used. An example of measurements made with a throttling calorimeter is given in the next section.

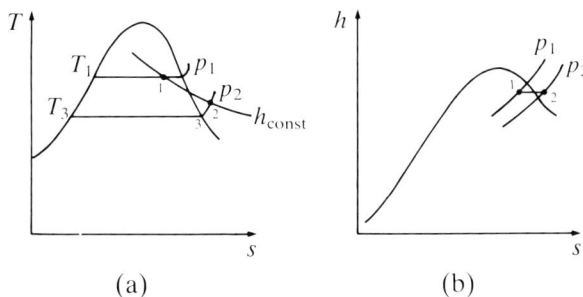

Fig. 6.2.5 Representation of throttling process:
(a) on T-s diagram
(b) on Mollier diagram

6.3 Experiment 7: Performance of a Steam Power Plant

Experience in the use of tables and diagrams of thermodynamic properties in the representation of thermodynamic processes is essential in order to consolidate an understanding of this important area of the subject. A practical difficulty arises in that the most instructive example for this purpose is undoubtedly the analysis of the performance of a steam power plant. The traditional laboratory steam power plant incorporates a large steam engine, a machine that is almost completely obsolete and, with the associated boiler and condenser, occupies a great deal of space. The alternative is a small steam turbine, which approaches current practice more closely but is also bulky and expensive and in addition has an efficiency that is very low when compared with that of the large steam turbines that are used in industrial applications.

The authors' solution to this problem is to make use of a very small steam power plant, Fig. 6.3.1, which, while occupying little space, nevertheless permits investigations of a real steam cycle and gives practice in the use of *T-s* and *h-s* diagrams. The efficiency of the unit is necessarily low, but this does not invalidate the principles involved.

The plant is shown schematically in Fig. 6.3.2. The source of energy is two electrical

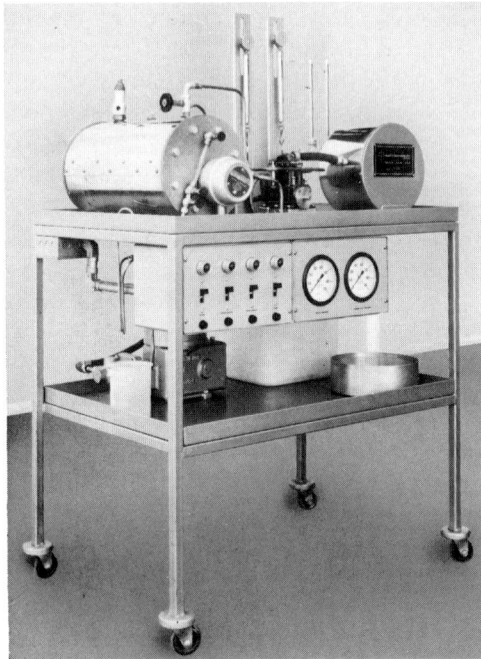

Fig. 6.3.1 Miniature steam power plant

immersion heaters of 3 kW capacity, and the boiler supplies unsuperheated steam to the steam motor, Fig. 6.3.3, a small trunk piston machine of 25·4 mm bore and stroke running at about 2000 rev/min.

The power is absorbed by a small rope brake and the engine exhausts to a condenser operating at atmospheric pressure. Condensate flow is measured by collection and weighing over a known time-interval.

Water is supplied to the boiler by an electrically driven feed pump. The experimen-

126

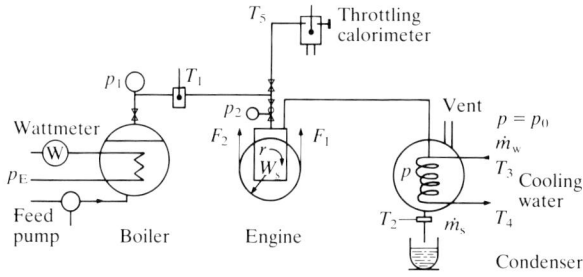

Fig. **6.3.2** Flow circuit of steam power plant

tal procedure is to run the unit until steady-state conditions are achieved, then to switch off the boiler feed pump and take a round of observations from which an energy balance may be derived and the Rankine cycle efficiency calculated. Under these conditions the electrical power input to the boiler may be equated with the latent heat imparted to the steam, since no feed water is entering the boiler.

The cycle is represented on the T-s diagram in Fig. 6.3.4, on which the state points may be compared with those shown in Fig. 6.2.2 for the Rankine cycle without superheat.

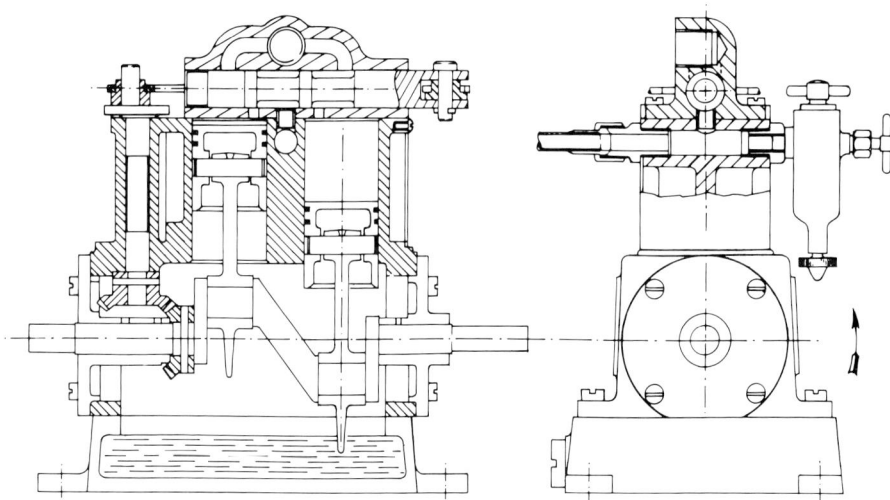

Fig. **6.3.3** Miniature steam motor

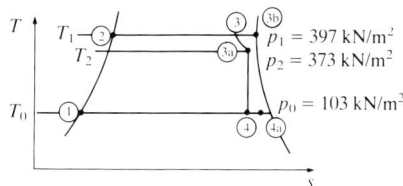

Fig. **6.3.4** Temperature–entropy diagram, small steam power plant

127

6.3.1 Measurements and Calculations

A typical set of observations is summarized below:

Barometric pressure	$P_0 = 103 \text{ kN/m}^2$
Boiler pressure, gauge	294 kN/m^2
	$p_1 = 397 \text{ kN/m}^2$
Stop-valve pressure, gauge	270 kN/m^2
	$p_2 = 373 \text{ kN/m}^2$
Boiler steam temperature	$T_1 = 415 \cdot 2 \text{K}$
Condensate temperature	$= 340 \text{K}$
Steam flow rate	$\dot{m}_s = 0 \cdot 002\,23 \text{ kg/s}$
Steam motor power output	$W_s = 136 \text{W}$
Electrical power input to boiler	$P_E = 5430 \text{W}$

A separate determination, described below, shows that for the steam entering the motor the dryness faction $x = 0 \cdot 98$.

The corresponding values of enthalpy and entropy may be read from a large-scale T-s or h-s (Mollier) diagram or from steam tables [18]:

State point (1), water at atmospheric pressure and boiling temperature	$P_0 = 103 \text{ kN/m}^2$ $T_0 = 373 \cdot 4 \text{K}$ $h_c = 419 \cdot 6 \text{ kJ/kg}$
State point (2), water at boiler pressure and boiling temperature	$p_1 = 397 \text{ kN/m}^2$ $T_1 = 416 \cdot 3 \text{K}$ $h = 603 \cdot 1 \text{ kJ/kg}$
State point (3b), dry saturated steam at boiler pressure (from steam tables; observed stop-valve temperature 415·2K)	$p_1 = 397 \text{ kN/m}^2$ $T_1 = 416 \cdot 3 \text{K}$

$$h = 2737 \cdot 2 \text{ kJ/kg}$$
$$s = 6 \cdot 90 \text{ kJ/kg K}$$

State point (3), steam at boiler pressure, $x = 0 \cdot 98$	$p_1 = 397 \text{ kN/m}^2$ $T_1 = 416 \cdot 3 \text{ K}$

$h = 603 \cdot 1 + 0 \cdot 98 \, (2737 \cdot 2 - 603 \cdot 1) \qquad = 2694 \cdot 5 \text{ kJ/kg}$
$s = 1 \cdot 77 + 0 \cdot 98 \, (6 \cdot 90 - 1 \cdot 77) \qquad\quad = 6 \cdot 80 \text{ kJ/kg K}$

State point (3a), steam at stop-valve pressure after throttling process from state point (3)	$p_2 = 373 \text{ kN/m}^2$ $T_2 = 414 \cdot 1 \text{K}$ $h = 2694 \cdot 5 \text{ kJ/kg}$
State point (4), steam after isentropic expansion from state point (3a)	$P_0 = 103 \text{ kN/m}^2$ $T_0 = 373 \cdot 4 \text{K}$ $s = 6 \cdot 85 \text{ kJ/kg}$ $h = 2477 \cdot 9 \text{ kJ/kg}$

When conditions at all the state points have been calculated, we can determine the Rankine efficiency of the cycle.

Heat imparted to the working fluid between state points (1) and (3) or (3a):

$$2694 \cdot 5 - 419 \cdot 6 = 2274 \cdot 9 \text{ kJ/kg.}$$

Work output ("adiabatic heat drop") between state points (3a) and (4):

$$2694 \cdot 5 - 2477 \cdot 9 = 216 \cdot 6 \text{ kJ/kg}$$

Rankine (thermal) efficiency $\dfrac{216 \cdot 6}{2274 \cdot 9} = 9 \cdot 5$ percent.

If the efficiency of the experimental plant equalled that of the Rankine cycle, the power output would equal the product of the steam flow rate and the adiabatic heat drop:

$$W_{s(Rankine)} = 0 \cdot 002 \, 23 \times 216 \cdot 6 \times 10^3 = 483 \text{W}.$$

The efficiency of the steam motor relative to the Rankine efficiency is thus:

$$\frac{W_s}{W_{s(Rankine)}} = \frac{136}{483} = 28 \cdot 1 \text{ percent.}$$

The "heat drop" between steam motor inlet and exhaust (i.e. the reduction in enthalpy) may be taken as

$$216 \cdot 6 \times 0 \cdot 281 = 60 \cdot 9 \text{ kJ/kg.}$$

6.3.2 Discussion of Results

The process taking place in the steam motor is an example of an expansion with friction and reheat. It is discussed in detail in Chapter 9. Here it will only be remarked that one way of visualizing the process is to consider it as taking place in two stages: an ideal frictionless adiabatic expansion from state (3a) to state (4) (as in a Rankine engine), followed by dissipation of a proportion of the displacement work performed in the course of this expansion as heat, which is returned to the working fluid in the course of a transition from state (4) to state (4a).

This interpretation is clearly reasonable. The authors' steam motor is of very small size and the mechanical efficiency is low, ~ 50 percent. It follows that at least half the work performed by the steam on the pistons is dissipated as friction, and much of this frictional heat will be returned to the steam in its passage through the motor.

We may readily calculate the theoretical heat input to the steam as the product of the mass flow rate and the change in enthalpy from state point (2) to state point (3):

$$P_s = 0 \cdot 002 \, 23 \, (2694 \cdot 5 - 603 \cdot 1) \times 10^3 = 4664 \text{W}$$

whence efficiency of boiler

$$\frac{P_s}{P_E} = \frac{4664}{5430} = 85 \cdot 9 \text{ percent}$$

Finally, the overall thermal efficiency of the plant is given by:

$$\eta_{th} = \frac{W_s}{P_E} = \frac{136}{5430} = 2 \cdot 5 \text{ percent}$$

compared with a Rankine efficiency of $9 \cdot 5$ percent.

6.3.3 Further Experiments and Questions

(a) Carry out a complete First Law analysis of the plant and draw up an energy balance.

(b) Study the characteristics of the steam motor by measuring the power output at constant speed and varying steam pressure. Estimate the mechanical efficiency by the "Willans Line" method (p. 197)—see Fig. 6.3.5.

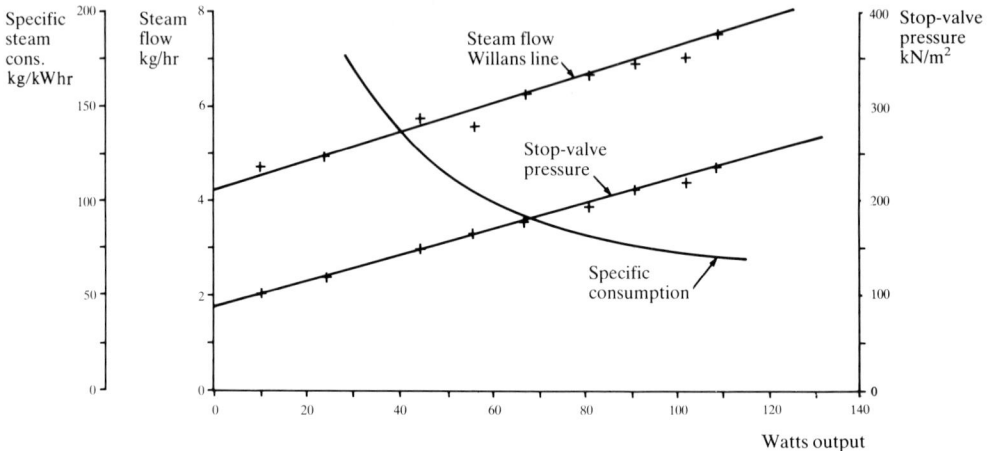

Fig. 6.3.5 Performance of small steam motor

(c) Employ the boiler to determine the relation between the pressure and temperature of saturated steam (the "Marcet boiler" experiment).

6.3.4 Determination of Dryness Fraction by the Throttling Calorimeter

Referring to Fig. 6.2.5 (p. 125), readings made by the authors in conjunction with the test results analysed above were as follows:

p_1	397 kN/m²
T_1	415·2K
p_2	103 kN/m²
T_2	382·5K

Saturated steam at pressure p_2 has a temperature $T_2 = 373·4$K, state point (3). The steam after expansion in the throttling calorimeter is thus superheated by 9·1°C. The corresponding enthalpy, determined from steam tables, is:

$$h = 2694·5 \text{ kJ/kg}$$

6.4 The Vapour Compression Refrigeration Cycle

In Chapter 5 we have discussed the heat pump and refrigerator and observed that ideally they would operate on a reversed Carnot cycle. However, just as, in the case of steam plant, the Carnot cycle is replaced by the Rankine cycle, the practical vapour compression cycle is a modification of the reversed Carnot cycle. Fig. 6.4.1 shows the vapour compression refrigeration circuit and the corresponding temperature–entropy diagram. The successive phases of the cycle are as follows:

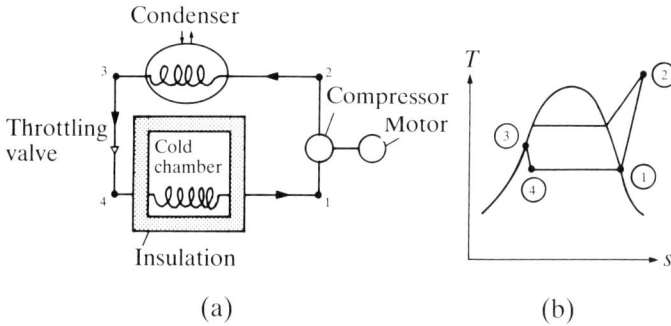

Fig. 6.4.1 (a) Vapour compression refrigeration circuit
(b) Representation on *T-s* diagram

1–2 Compression of working fluid, usually by a piston compressor.

2–3 Cooling of the working fluid with heat transfer to the surroundings. The fluid normally leaves the compressor in a superheated condition and is first cooled to the saturation temperature corresponding to the compressor delivery pressure. It is then condensed with transfer of the latent heat of vaporization and, since it is desirable that the working fluid should leave the condenser wholly in liquid form, the working conditions are so designed that the condensed liquid is further cooled to slightly below the condensation temperature.

3–4 Expansion of the condensed liquid through a throttling valve with corresponding fall in pressure. The process takes place at constant enthalpy and some evolution of vapour takes place. This phase represents a departure from the Carnot cycle analogous to that represented by the boiler feed pump in the Rankine cycle.

4–1 The working fluid is circulated through the space from which heat is to be extracted in the course of which it vaporizes, taking up the latent heat of vaporization in the process. As it is undesirable for mechanical reasons that working fluid should enter the compressor in liquid form, the system is so proportioned that the vapour leaves the evaporator in a slightly superheated condition.

It will be apparent that water could, in principle, be used as the working fluid in a vapour compression refrigeration cycle, but only between temperature limits substantially above the freezing point of water. A practical working fluid (refrigerant) must meet the requirement that its saturation pressure should not be too high at the temperature prevailing in the condenser, or too low at the temperature in the evaporator. If the vapour pressure is less than atmospheric at any point in the cycle, leakage of atmospheric air into the refrigerant circuit can take place, with consequent disturbance to the working cycle and possible icing-up of the throttle valve. Carbon dioxide (CO_2), sulphur dioxide (SO_2) and ammonia (NH_3) have all been widely used as working fluids in the past; they have now been largely replaced by a range of fluorine compounds that are less toxic and have better thermodynamic properties. Refrigerant R22 (CHF_2Cl), the refrigerant used in the heat pump of Experiment 6, is an example of this range of refrigerants.

6.4.1   Experiment 6: Further Analysis

While in Fig. 6.4.1 the vapour compression cycle has been represented on the *T-s* diagram, it is more usual and convenient to represent the cycle on a diagram having as axes pressure and enthalpy. A series of pressure–enthalpy diagrams for commercial refrigerants is given in [20] and the diagram for refrigerant R22 is reproduced to a reduced scale in Fig. 6.4.2. The diagram is particularly convenient for representing the cycle because all the relevant state points are located either at the condenser pressure or at the evaporator pressure. The compressor and the throttling valve operate between these two pressures, while pressure differences across the condenser and the evaporator are negligible.

The refrigerator incorporated in the heat pump of Experiment 6 was fitted with thermocouples at the point indicated in Fig. 5.5.2 (p. 111), and Table 6.4.1 shows two sets of readings, one with the apparatus functioning as a heat pump and one functioning as an air conditioner. In the latter case the positions of the evaporator and condenser in the circuit are interchanged by a system of solenoid-operated valves. Evaporation then takes place in the heat exchanger located in the air duct, resulting in cooling of the air, while condensation takes place in the water-cooled heat exchanger, with heat rejection to the circulating water.

Table 6.4.1   Heat Pump Refrigerant Temperatures

	Heat pump	Cooler
Compressor inlet T_5	288K	279K
Compressor discharge T_6	353K	330K
Condenser outlet T_7	334K	313K
Evaporator inlet T_8	288K	279K

The corresponding cycles are plotted on the pressure–enthalpy diagram in Fig. 6.4.2. It is assumed that the temperature T_7 of the liquid refrigerant leaving the condenser corresponds to the saturation temperature of the fluid at condenser pressure. This cannot be proved in the absence of a direct pressure measurement, but any possible error is small. State point 3 must lie on the saturated liquid line and is assumed to correspond to the pressure in the condenser.

The throttling process 3–4 is represented by a vertical line on the diagram (constant enthalpy), and since for both sets of readings $T_5 = T_8$ we may assume in the present case that state point 1 lies on the isobar corresponding to the evaporator pressure and on the saturated vapour line. It is noteworthy that a straight line joining state points (1) and (2) on the diagram, representing the compression process, in each case almost

132

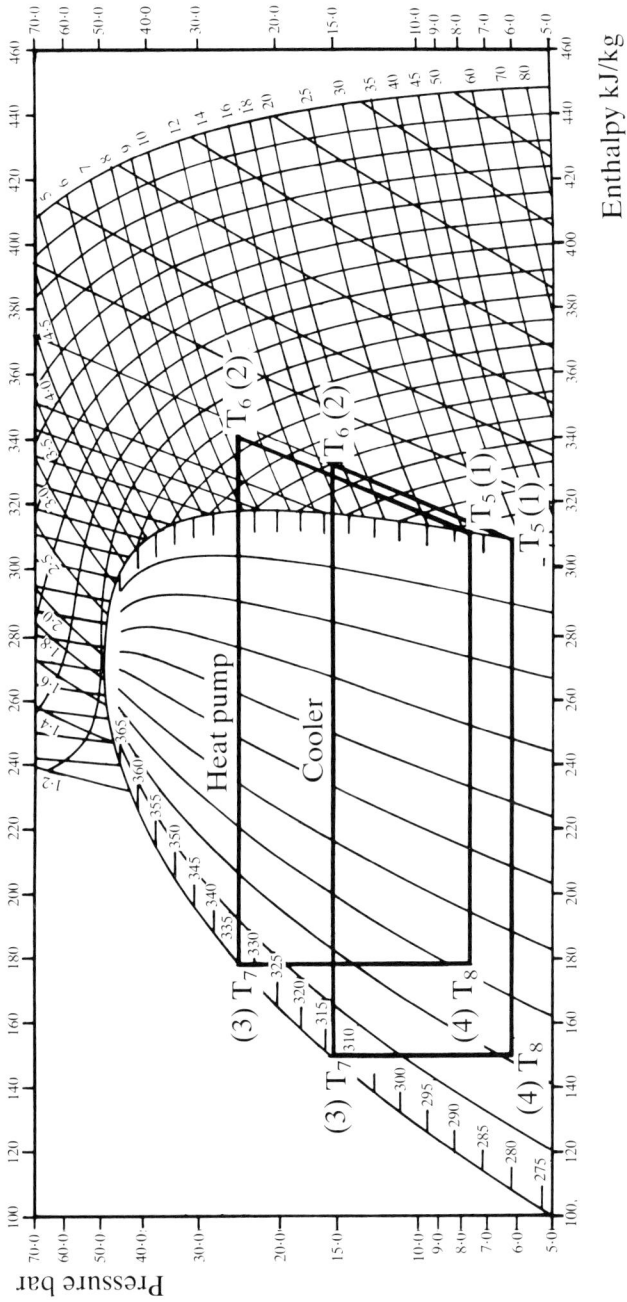

Fig. 6.4.2 Pressure–enthalpy diagram for refrigerant R22

133

coincides with a parameter line of constant entropy; the compression process is approximately isentropic.

If the rate of refrigerant flow is also measured (not the case in the authors' tests), the diagram Fig. 6.4.2 may also be used for calculations of energy flows in the cycle.

7

Gaseous Mixtures; Psychrometry

7.1 The Gibbs–Dalton Law of Gaseous Mixtures

In thermodynamics we are frequently concerned with the properties of gaseous mixtures. Common examples are the products of combustion in an internal combustion engine and air itself. The Gibbs–Dalton Law, which was established empirically, enables us to predict the properties of a gaseous mixture from those of its constituents, provided that there is no chemical reaction between them.

Consider a volume V occupied by a mixture of different gases at pressure p and temperature T. The Gibbs–Dalton Law states that the pressure exerted by each constituent gas and the value of its internal energy are each equal to that of the gas if it alone occupied the volume V at the same temperature. (Dalton's original law of partial pressures was later extended to include reference to internal energies.) It is easily shown that the enthalpy of the mixture also represents the sum of the individual enthalpies.

We may thus write the total pressure of a gaseous mixture as the sum of the partial pressures:

$$p = p_1 + p_2 + p_3 + p_{n...} = \Sigma p_n \qquad (7.1)$$

The additive nature of partial pressures is consistent with the implications of the kinetic theory, which states that gaseous pressure is the consequence of molecular impacts and also assigns the same kinetic energy to the molecules of all gases at any specific temperature. Avogadro's Hypothesis indicates that all gases at the same conditions of temperature and pressure contain the same number of molecules per unit volume, and we may conclude that the volume V contains the same molecular population for any particular values of p and T, the molecules having identical mean kinetic energies irrespective of the composition of the gas concerned.

If we assume that the mixture is made up of perfect gases, we may deduce a number of simple relationships which find practical application in the field of psychrometry and in the study of combustion processes.

The total mass m of the mixture is given by:

$$m = m_1 + m_2 + m_3 + \ldots m_n = \Sigma m_n \qquad (7.2)$$

For each constituent gas we may write:

$$p_n V = m_n R_n T = p V_n \qquad (7.3)$$

where V_n is the volume that the gas would occupy if compressed to the pressure p of the mixture. It follows from equation (7.3) that

$$\frac{V_n}{V} = \frac{p_n}{p} \tag{7.4}$$

and

$$V = V_1 + V_2 + V_3 + \ldots V_n = \Sigma V \tag{7.5}$$

The partial pressures are thus in the same proportion as the partial volumes would be if each gas were compressed isothermally to the full pressure of the mixture.

We can calculate various properties of the mixture as follows:

$$\rho = \frac{\dot{m}}{V} = \frac{V_1\rho_1 + V_2\rho_2 + V_3\rho_3 + \ldots}{V} \tag{7.6a}$$

where ρ_1, ρ_2, $\rho_3 \ldots$ are the densities of the constituent gases at pressure p and temperature T.

$$R = \frac{m_1 R_1 + m_2 R_2 + m_3 R_3 + \ldots}{m} \tag{7.6b}$$

$$c_v = \frac{m_1 c_{v1} + m_2 c_{v2} + m_3 c_{v3} + \ldots}{m} \tag{7.6c}$$

$$c_p = \frac{m_1 c_{p1} + m_2 c_{p2} + m_3 c_{p3} + \ldots}{m} \tag{7.6d}$$

Molecular weight:

$$M = \frac{V_1 M_1 + V_2 M_2 + V_3 M_3 + \ldots}{V} \tag{7.6e}$$

Air is a mixture of gases, by volume:

$$
\begin{array}{ll}
N_2 & 78\% \\
O_2 & 21\% \\
Ar & 1\%
\end{array}
$$

We can calculate the gas constant R and an equivalent "molecular weight" for air from a knowledge of the properties of the constituent gases.

7.2 Psychrometry

The atmosphere is a mixture of dry air and water vapour. The properties of moist air are of importance in connection with air conditioning, evaporative cooling—as in the cooling towers of power stations—and, of course, in meteorology. The study of the properties of moist air is known as psychrometry.

The theoretical basis of the subject is built up on the assumption that a mixture of dry air and water vapour in the proportions encountered in the atmosphere may be treated at a mixture of perfect gases, thus allowing the methods outlined in the last section to be applied. We thus assume that both air and water vapour obey an equation of state of the form $pV = RT$.

For air:

$$p_a V_a = R_a T$$
$$R_a = 0.2871 \text{ kJ/kgK} \tag{7.7}$$

For water vapour:

$$p_s V_s = R_s T$$
$$R_s = 0.4615 \text{ kJ/kgK}$$

Fig. 7.2.1 shows the correction factors that must be applied to the value of R in each case to compensate for the deviations of the real gas or vapour from the ideal equation. It is obvious, particularly in view of the fact that the proportion of water vapour in moist air very rarely exceeds 3 percent, and that we are only concerned with pressures in the neighbourhood of atmospheric, that we may ignore these deviations.

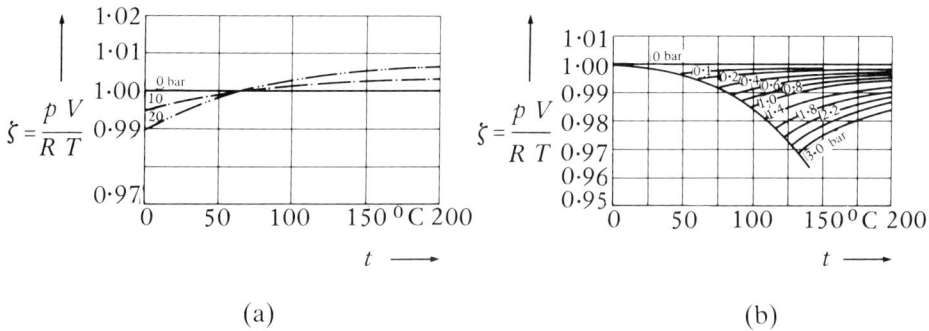

Fig. 7.2.1 (a) Correction factor applicable to perfect gas equation for air
(b) Correction factor to perfect gas equation for water vapour

Several special terms used in psychrometry require definition. The specific humidity or moisture content ω is the ratio of the mass of water vapour m_s to the mass of dry air m_a in any given volume of moist air:

$$\omega = \frac{m_s}{m_a} \tag{7.8}$$

Specific humidity is dimensionless and its value can range from $\omega = 0$ (dry air) to $\omega = \infty$ (pure water vapour). The area of practical interest ranges from $\omega = 0$ to about $\omega = 0.1$.

Since we are treating water vapour as a perfect gas we can define a relationship between ω and the partial pressure p_s of the water vapour in the mixture:

$$p_s V = m_s R_s T$$
$$p_a V = m_a R_a T$$

$$\therefore \omega = \frac{m_s}{m_a} = \frac{p_s}{p_a} \cdot \frac{R_s}{R_a} = 0\cdot662 \frac{p_s}{p-p_s} \tag{7.9}$$

where $p = p_a + p_s$.

The partial pressure of water vapour in moist air cannot exceed the pressure p_g at which, in the absence of air, water would boil at the prevailing temperature (the boiling point of water at low pressures is very much depressed; at a pressure of 8 mmHg it is only 7°C). When the partial pressure reaches this value the air is said to be saturated, and any additional moisture will either be deposited as dew or frost or will be present as a fog or cloud.

The relative humidity φ is defined as:

$$\varphi = \frac{p_s}{p_g} \tag{7.10}$$

Since we may write

$$p_s V = m_s R_s T$$
$$p_g V = m_g R_g T$$

then

$$\varphi = \frac{m_s}{m_g}$$

The relative humidity φ is thus the ratio of the mass of water vapour actually present to the mass that would be present if the air were saturated at the same conditions of temperature and pressure. φ, which is often expressed as a percentage, is of greater practical significance in air conditioning than the specific humidity, since the degree of comfort corresponding to any given atmospheric condition depends much more on the relative humidity than on the actual moisture content. Drying processes are also dependent on the relative humidity rather than on the specific humidity.

The dew-point is the temperature to which unsaturated air must be cooled at constant pressure for it to become saturated. If the moisture content at temperature T is m_s, equation (7.9) indicates that if we cool the moist air at constant pressure, the partial pressure p_s of the water vapour must remain constant. The dew-point is reached when the pressure p_s equals the saturation pressure p_{gd} corresponding to the dew-point temperature, Fig. 7.2.2. The dew-point is commonly determined by cooling a metal plate having a bright surface and noting the temperature at which dew is first deposited on it. The relative humidity may then be found from equation (7.10).

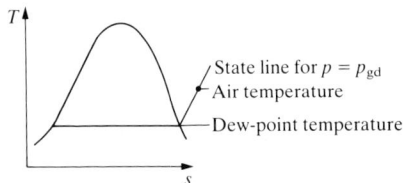

Fig. 7.2.2 Representation of the dew-point on the *T-s* diagram

As an example, assume that the air temperature is 20°C and the dew-point occurs at 10°C. The relative saturation pressures are:

$$t = 20°C \qquad p_g = 2360 \text{ N/m}^2$$
$$t = 10°C \qquad p_{gd} = 1230 \text{ N/m}^2$$
$$\varphi = \frac{1230}{2360} = 52 \text{ percent}$$

Here p_g = saturation pressure at the original temperature T.

The simplest and most widely used method of determining the relative humidity is by means of a wet- and dry-bulb thermometer. If unsaturated air flows past a thermometer having a wetted sleeve of cotton around the bulb, the temperature recorded will be less than the actual temperature of the air, owing to evaporation from the wetted sleeve. Within a wide range of air velocities, approximately from 2 m/s to 40 m/s, the equilibrium temperature registered by the thermometer remains constant and is a function of the relative humidity of the air. A common form of this instrument is the sling or whirling hygrometer, which consists of two glass thermometers, one having its bulb surrounded by a sleeve wetted from a small reservoir. The difference between the wet-bulb and the dry-bulb temperatures is a measure of the relative humidity. Under saturated conditions the temperatures are identical, and the depression of the wet-bulb reading increases with increasing dryness. The psychrometric chart, now to be described, permits an assessment of the moisture content of the air from a knowledge of the wet- and dry-bulb temperatures.

7.3 The Psychrometric Chart

The specific enthalpy of moist air at atmospheric pressure may be expressed by a simple approximate equation. It is usual in psychrometric calculations to take $t = 0°C$ and $\omega = 0$, or dry air, as corresponding to zero enthalpy. Then for 1 kg of dry air containing ω kg of moisture we may write:

$$h_{a+s} = c_{pa} \cdot t + \omega(c_{ps} \cdot t + L) \tag{7.11}$$

where

c_{pa} = specific heat of air at constant pressure
c_{ps} = specific heat of water
L = latent heat of water at temperature t.

As a good approximation we take $c_{pa} = 1$ kJ/kgK, take $L = 2500$ kJ/kg (the value at 0°C), and adjust the value of c_{ps} to take into account the fall in L with increasing temperature. We may then write:

$$h_{a+s} = t + \omega(1 \cdot 86t + 2500) \text{ kJ/kg} \tag{7.12}$$

If the air is super-saturated and contains liquid water in the form of mist or droplets, $\omega > \omega_s$, the value corresponding to saturated conditions, we must add a further term:

$$h_{a+s} = t + \omega_s(1 \cdot 86t + 2500) + 4 \cdot 19(\omega - \omega_s) \tag{7.12a}$$

Equation (7.12) can be used to draw up a chart to facilitate psychrometric calculations. Several different forms of this chart are in use, and English practice differs from that on the Continent of Europe. Fig. 7.3.1 shows schematically a chart of h_{a+s} against t. The enthalpy of dry air is represented by a straight line through the origin, while individual values of ω are represented by a series of parallel lines. With the aid of steam tables which tell us the pressure at which water boils corresponding to any given temperature, and making use of equation (7.9) which links ω with the partial pressure we may draw in a curve corresponding to $\varphi = 1$ or saturated conditions. Curves for intermediate values of φ may be interpolated directly.

The value of the chart is greatly increased if in addition we plot parameter lines corresponding to different wet-bulb temperatures, using data that must be obtained experimentally. Two such lines are shown in Fig. 7.3.1. The line corresponding to a wet-bulb temperature of 10°C must by definition intersect the saturation line at a point corresponding to $t = 10$°C, since, as we have already seen, under saturated conditions, the wet-bulb temperature is equal to that of the air. Experiment shows that for completely dry air a wet-bulb temperature of 10°C corresponds to an air temperature of about 28·2°C. The wet-bulb temperature 10°C line thus runs from the point on the saturation line corresponding to $t = 10$°C to a point on the dry air line corresponding to $t = 28·2$°C. The line is very nearly straight between these points.

It will thus be apparent that if for any given atmospheric condition we know the wet-bulb temperature and the temperature t (the dry-bulb temperature) we may locate a point on the wet-bulb parameter line and read off the corresponding values of ω and φ.

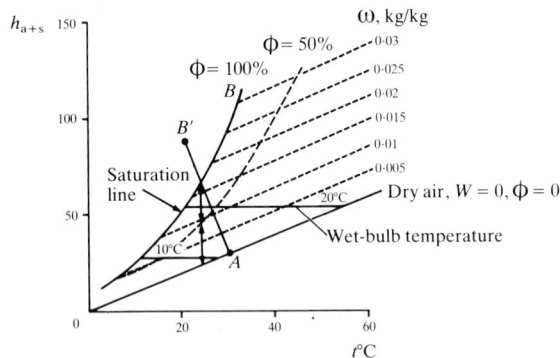

Fig. 7.3.1   Enthalpy–temperature chart for moist air

A diagram in the form of Fig. 7.3.1 is inconvenient for practical use because the area of interest is concentrated in a narrow zone between the $\varphi = 0$ and $\varphi = 1$ lines, while the various parameter lines intersect at acute angles, making the accurate location of points on the chart difficult.

This difficulty is overcome in the case of the psychrometric chart published by the Institution of Heating and Ventilating Engineers [21], reproduced in Fig. 7.3.2, by replotting the diagram as follows. Point A, Fig. 7.3.1, corresponds to dry air at 30°C and point B to saturated air at the same temperature. Line AB is rotated into position AB', where AB' is normal to the parameter line for dry air. AB' now represents $t = 30$

140

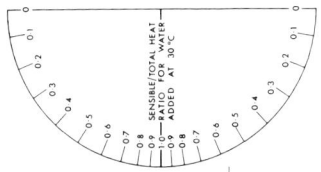

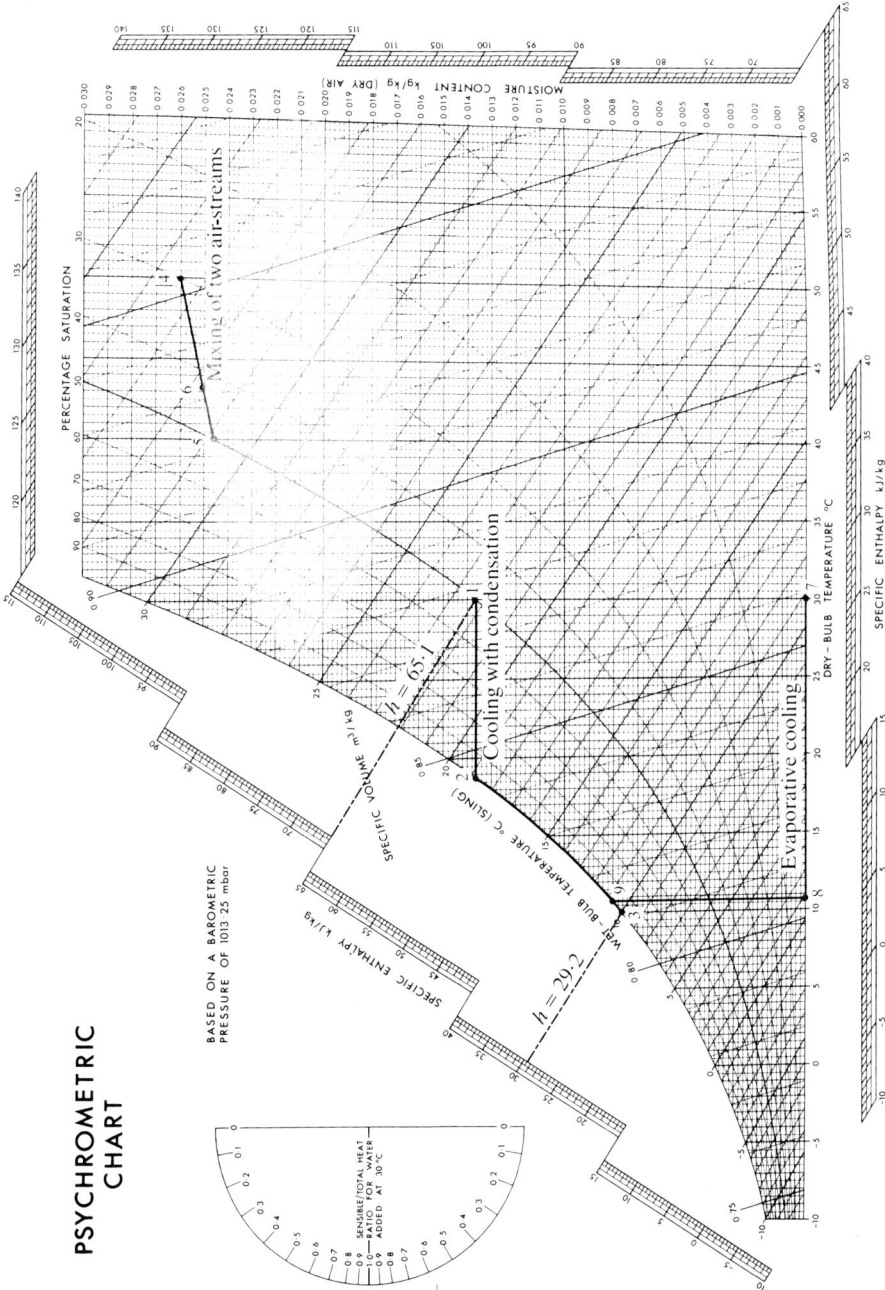

Fig. 7.3.2 The I.H.V.E. psychrometric chart

141

on the modified diagram, and it will be appreciated that the interval between the curves for $\varphi=0$ and $\varphi=1$ has been greatly increased.

Points corresponding to other values of t over the full range are treated accordingly, although the geometry is such that the lines of constant temperature are not quite parallel but diverge slightly; this is of no disadvantage in using the diagram. In the I.H.V.E. chart the abscissa corresponds to the line for dry air in Fig. 7.3.1, and it is graduated for a range of dry-bulb temperatures from $-10°C$ to $60°C$. The inclined ordinate shows the specific enthalpy, while lines of constant moisture content ω lie parallel to the temperature axis. The chart also shows curves of φ from 10 percent to saturation and lines for wet-bulb temperature at intervals of $1°C$ from $-12°C$ to $36°C$. Additionally curves are shown for the specific volume of moist air in m^3/kg.

Although the psychrometric chart is drawn for a standard atmospheric pressure of 760 mmHg it may be used without appreciable error for atmospheric pressures within the range ±10 percent of this value. Several typical processes will now be described to illustrate the use of the chart.

(a) Cooling of Moist Air with Condensation of Moisture

Let the initial condition of the air be:

Dry-bulb temperature   30°C
Wet-bulb temperature   22·3°C

This is plotted as point 1 on Fig. 7.3.2, from which we may deduce the following:

$$\varphi = 50 \text{ percent}$$
$$\omega = 0\cdot0137 \text{ kg/kg}$$

The cooling process is represented by a horizontal line on the chart, ω remaining constant until the saturation line is reached at the dew-point, point 2:

dry-bulb temperature $= 18\cdot7°C$
wet-bulb temperature $= 18\cdot7°C$

If the air is further cooled to a temperature of 10°C, point 3, moisture will be deposited in liquid form and we arrive at the conditions:

dry-bulb temperature $= 10°C$
wet-bulb temperature $= 10°C$
$\omega = 0\cdot0077 \text{ kg/kg}.$

The moisture content has been reduced by $0\cdot0137 - 0\cdot0077 = 0\cdot0060 \text{ kg/kg}$. The specific enthalpy has fallen from 65·1 kJ/kg at point 1 to 29·2 kJ/kg at point 3, and this enables us to calculate the heat that must be removed to bring about this degree of cooling.

(b) Mixing of Airstreams

Consider two airstreams with mass flow rates of the dry air component of each stream respectively $\dot{m}_{a1}$ and $\dot{m}_{a2}$, and moisture contents ω_1 and ω_2. We may write the following relationships to define the conditions after mixing has taken place:

$$\dot{m}_{a1} + \dot{m}_{a2} = \dot{m}_a$$

$$\omega_1 \dot{m}_{a1} + \omega_2 \dot{m}_{a2} = \omega \dot{m}_a$$

$$h_{(a+s)1} \dot{m}_{a1} + h_{(a+s)2} \dot{m}_{a2} = h_{a+s} \dot{m}_a$$

As these equations are linear, the mixing process may be represented by plotting points corresponding to the conditions in each airstream on the psychrometric chart, joining them by a straight line, and dividing this line in inverse ratio to the relative mass flow rates.

Fig. 7.3.2 represents the mixing of unit mass of air at 50°C, $\varphi = 30$ percent, point 4, and twice unit mass at 40°C, $\varphi = 50$ percent, point 5. The mixture is represented by point 6: temperature 43·3°C, $\varphi = 42$ percent.

This suggests a possible situation: that in which two airstreams, both unsaturated, give rise on mixing to a situation of over-saturation, leading to the deposit of moisture. This is the explanation of the commonly observed phenomenon that breath is visible on a cold day. Breath leaving the lungs has a temperature of about 36°C and is very nearly saturated. These conditions are beyond the limits of the chart, but it is obvious that such an airstream, when mixed with air at temperatures in the region of freezing point, will give rise to over-saturation.

(c) Evaporative Cooling

This phenomenon is most easily explained in terms of the wet-bulb thermometer, already described. Suppose a stream of air at temperature t_1 to impinge on the thermometer bulb which is surrounded by a wick saturated in water, and assume that the process has been going on long enough for equilibrium conditions to have been established. Neglecting heat transfer from the surroundings, the process is essentially one of mixing between the water that is being drawn up by the wick and some part of the air flowing past the apparatus.

While it could not be foreseen on theoretical grounds, experiment shows that the relation between the temperatures t_1 and t_2 is consistent with the establishment of complete saturation of some of the air passing the bulb. The process can best be described in terms of the psychrometric chart and a specific set of conditions, Fig. 7.3.2.

Suppose that the air approaching the wet bulb is at a temperature of 30°C and completely dry, point 7. Consider only that portion of the air that becomes fully saturated in its passage past the bulb. We are concerned with the mixing of dry air at 30°C and an appropriate quantity of water at the wet-bulb temperature. The enthalpy of the moist air leaving the bulb must then equal the sum of the enthalpies of the dry air and the water that is subsequently evaporated. The process is one of energy exchange with the air losing sensible heat which is expended in vaporizing the water carried away by the airstream. The process may be envisaged as taking place in two stages. Imagine the air to be cooled from point 7 to point 8, the latter corresponding to the wet-bulb temperature. This heat is then transferred at constant temperature, in the case of our example 10·7°C, to the water and we arrive at point 9. The air is now saturated, $\omega = 0.008$, and at the wet-bulb temperature of 10·7°C.

The enthalpy of the dry air at point 7 is 30 kJ/kg and the enthalpy of the air plus its moisture content at point 9 is approximately 31 kJ/kg.

Wet-bulb temperatures have been established by experiment, and the results agree very nearly but not quite exactly with those predicted by the use of equation (7.12).

The analysis is identical in principle when the air approaching the wet-bulb is not initially dry. Strictly speaking, the mixing process applies only for fully turbulent conditions; this is the reason for the specification of a minimum air velocity of 2 m/s for the wet-bulb thermometer.

The fact that under adiabatic conditions both the air and water are cooled does not contradict the requirements of the Second Law. The increase in entropy of the water during evaporation is always greater than the reduction in entropy of the air during the cooling process.

While the above explanation of evaporative cooling sounds plausible, the phenomenon nevertheless has curious features. On the face of it, for instance, it contravenes the Zeroth Law of thermodynamics, p. 5. As was the case with the process of diffusion, the explanation is to be sought in the kinetic theory. The kinetic energy of individual molecules of a substance at a given temperature is distributed over a wide range, in accordance with a well-known statistical law, Fig. 7.3.3. The process of evaporative cooling implies that a certain number of water molecules possess sufficient kinetic energy to escape from the main body of water. In order to do so they must overcome the attractive forces of the surrounding molecules; these attractive forces correspond approximately to the latent heat of evaporation.

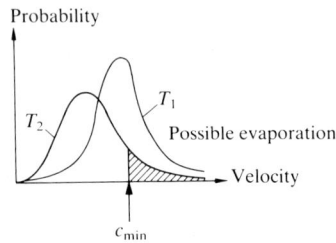

Fig. 7.3.3 Distribution of molecular kinetic energies

At 0°C this latent heat is more than twelve times as great as the mean kinetic energy of translation of the water molecules. The possibilities of escape are correspondingly small, and if in Fig. 7.3.3 the curve t_1 represents the distribution of kinetic energies of the air molecules and t_2 of the water molecules, only the proportion of the latter having kinetic energies lying within the cross-hatched area of the distribution curve have a chance of escaping from the water surface.

After escape from the water surface an H_2O molecule will have expended almost all its energy of translation, but subsequent impacts with O_2 and N_2 and occasionally with other H_2O molecules will compensate for this loss of energy and the molecule will rapidly achieve a mean kinetic energy corresponding to the wet-bulb temperature. The mean free path at atmospheric conditions is only in the neighbourhood of 10^{-4} mm, and many H_2O molecules will return to the water surface, experiencing a sharp increase in energy in the process. The macroscopic process of evaporative cooling corresponds to the excess of molecules escaping from the water surface over those returning to it. The heat transfer accompanying the process is brought about by

144

the circulation of H_2O molecules in this manner. Natural convection between the air and the water surface plays only a small part in the process.

If a higher degree of accuracy than is obtainable by the use of the psychrometric chart is required, psychrometric tables such as those produced in [22] may be employed. These give precise values of vapour pressure, relative humidity and dew-point over the whole range of wet- and dry-bulb temperatures that are of practical interest. Detailed recommendations regarding the correct use of psychrometers are given in this reference.

7.4 Experiment 8: Air-Conditioning Apparatus

The process of air conditioning, which is becoming of ever-increasing practical importance as the requirement for comfortable living and working conditions becomes more exacting, involves control of the temperature and humidity of the air. An air conditioner must thus have the capability of either warming or cooling the air and of increasing or decreasing its moisture content. Any air-conditioning apparatus must therefore include the following elements:

(a) A fan.
(b) A heater.
(c) A water or steam injector.
(d) Refrigerating equipment.
(e) Means for separating liquid water from the airstream.

Reduction of moisture content in the air is brought about by cooling the air to below the saturation temperature, extracting the condensed water, and if necessary re-heating the air to the desired temperature.

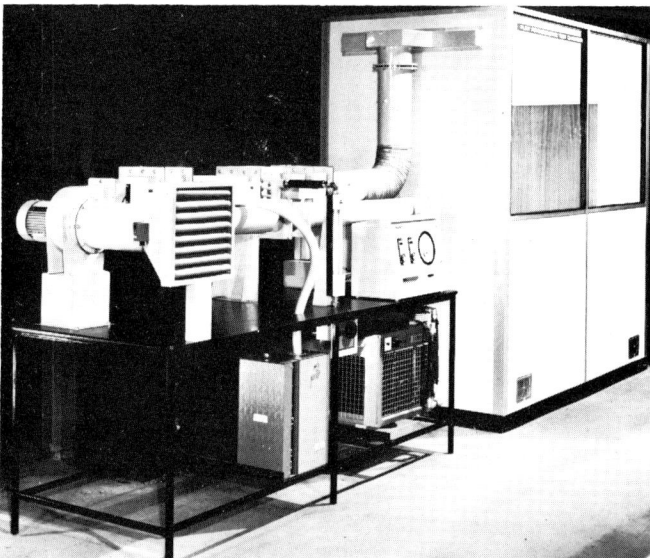

Fig. 7.4.1   Air-conditioning plant and psychrometric test chamber

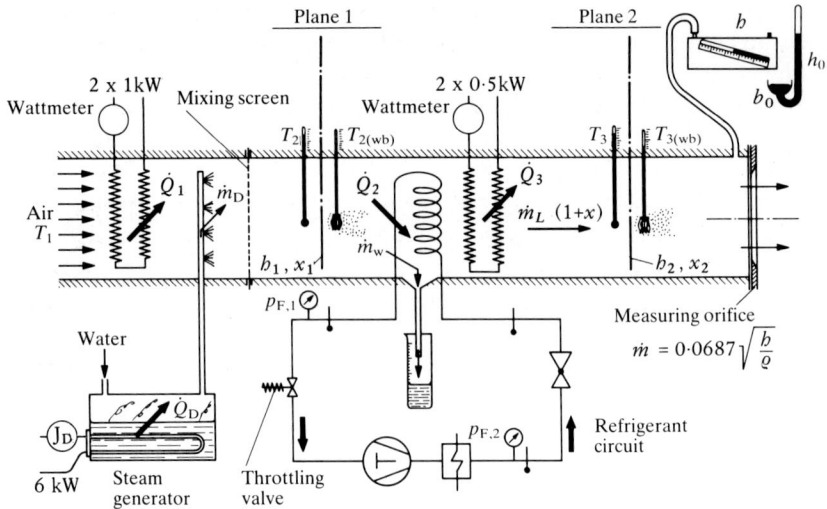

Fig. 7.4.2 Schematic arrangement of experimental air-conditioning apparatus

Fig. 7.4.1 shows an experimental apparatus used by the authors which incorporates all the necessary elements of an air-conditioning system, and Fig. 7.4.2 shows schematically the layout of the apparatus. It consists essentially of a rectangular duct through which air may be driven by a variable-speed fan. In the course of its passage through the duct the air encounters in succession:

(a) An electrical heating element.
(b) A steam injector.
(c) Screens to ensure a uniform velocity distribution and mixing of the injected steam with the air current.
(d) Wet- and dry-bulb thermometer.
(e) Cooling coil through which refrigerant is circulated and, in the same plane, a collector for moisture condensed from the airstream.
(f) A further heating element for warming the air after the drying process.
(g) A further wet- and dry-bulb thermometer.
(h) Sharp-edged orifice for flow measurement with accompanying manometer.

Associated with the air-conditioning duct itself are a refrigerator and an electrically heated boiler for producing steam that may be added to the airstream. The mass flow rate of air in the authors' apparatus was given by:

$$\dot{m} = 0 \cdot 0407 \sqrt{\frac{hh_0}{T_3}}$$

where h = head across measuring orifice, mmH_2O
          h_0 = barometer, mmHg
          T_3 = air temperature at orifice

This equation gives the rate of flow of moist air; it is very little affected by the level of relative humidity.

146

The rate of condensation of moisture from the airstream under cooling conditions is determined by collecting and measuring the condensate over a timed period.

7.4.1 Measurements and Calculations

A typical sequence of tests is described below and corresponding experimental observations are given in Table 7.4.1. These processes are represented on the psychrometric chart in Fig. 7.4.3.

Table 7.4.1 Experiment 8: Air-conditioning Apparatus: Observations

		Dry bulb	Wet bulb
Room air	t_2	19·8°C	15·9°C
Heated air	t_3	26·6°C	18·0°C
Steam injection	t_2	27·6°C	23·7°C
Cooled air	t_3	22·0°C	20·9°C
All tests	h	5·2 mmH$_2$O	
	$\dot{m}$	0·146 kg/s	

Ambient temperature   19·8°C
Barometer   740 mmHg

(a) Measurement of Temperature, Relative Humidity and Dew-point of Room Air

Table 7.4.1 shows:

$$T_2 = 292\cdot8\text{K}, \quad T_{2(wb)} = 288\cdot9\text{K}$$

The corresponding condition is plotted as point a in Fig. 7.4.3. Interpolating the values of wet- and dry-bulb temperature, we may read the following information directly from the chart:

Relative humidity $\varphi = 66$ percent
Moisture content $\omega = 9\cdot6$ gm/kg
Dew-point $= 13\cdot5$°C

(The dew-point is the saturation temperature corresponding to the observed moisture content.)

For completeness, in this and in all the subsequent tests, the wet- and dry-bulb temperatures of the air in the laboratory in the neighbourhood of the inlet to the apparatus should also be observed, using a whirling hygrometer.

(b) Addition of Heat to the Airstream

The authors' results are plotted as point b in Fig. 7.4.3. It will be observed that, as is to be expected, the moisture content and hence the dew-point of the air are unchanged but, as a consequence of the higher temperature, the relative humidity has fallen. The mass flow rate was observed in the authors' test and we may read off the specific enthalpies corresponding to points a and b from the psychrometric chart and calculate the corresponding heat input to the air:

147

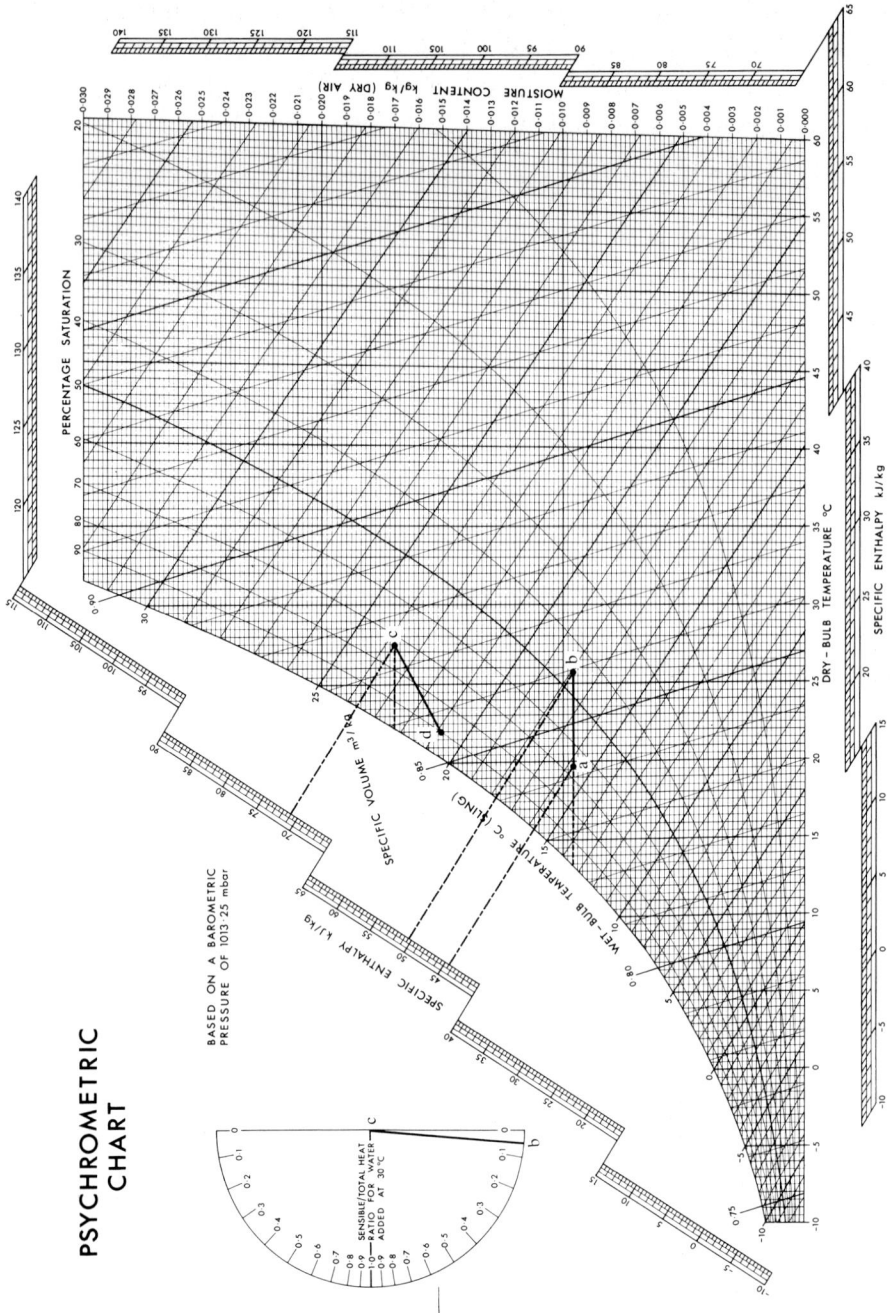

Fig. 7.4.3 Experimental results for air-conditioning apparatus: representation on the psychrometric chart

$$\dot{m} = 0 \cdot 146 \text{ kg/s}$$
$$h_b - h_a = 52 \cdot 6 - 45 \cdot 5 = 7 \cdot 1 \text{ kJ/kg}$$
$$\text{heat input} = 0 \cdot 146 \times 7 \cdot 1 = 1 \cdot 04 \text{kW}$$

This agrees well with the nominal 1 kW electrical input to the heating element.

(c)   Addition of Moisture to Airstream

The wet- and dry-bulb temperatures are plotted as point c and indicate the following conditions downstream of the steam injector:

Relative humidity φ = 72 percent
Moisture content ω = 16·9 gm/kg
Increase in moisture content   = 7·3 gm/kg
Dew-point   = 22·4°C
Increase in specific enthalpy point b to point $c = 20 \cdot 5$ kJ/kg

This experiment allows us to illustrate the purpose of the polar diagram given on the psychrometric chart. This is graduated to give a scale of values of the ratio of sensible to total heat for water added (nominally at 30°C but the scale is little changed when, as in the present case, the water is added at 100°C, the temperature of the steam generated in the boiler). The nomenclature of this polar diagram is perhaps slightly confusing. In the present case the sensible heat of the steam entering the airstream is approximately 100 kJ/kg, the latent heat approximately 2300 kJ/kg and the total heat 2400 kJ/kg. The ratio of sensible to total heat is thus approximately 0·04. A line drawn on the polar diagram parallel to that joining points b and c on the chart intersects the polar scale at the appropriate value, 0·04.

The purpose of the polar diagram is thus to permit prediction of the direction of the change of state consequent on the injection of steam of varying temperatures and degrees of dryness into the airstream.

The increase in enthalpy of the airstream between points b and c is:

$$0 \cdot 146 \times 20 \cdot 5 = 2 \cdot 99 \text{ kW}$$

The rate at which steam is added to the airstream between b and c is:

$$0 \cdot 146 \times 7 \cdot 3 = 1 \cdot 07 \text{ gm/s}$$

Taking the total heat of the steam as 2400 kJ/kg, the energy input from the steam is:

$$2400 \times 1 \cdot 07 \times 10^{-3} = 2 \cdot 57 \text{ kW}$$

This agrees fairly well with the increase in enthalpy deduced above.

(d)   Drying of Air by Cooling

If the relative humidity is initially sufficiently high, water condensed out of the airstream will begin to flow into the collecting vessel.

The results recorded in Table 7.4.1 are plotted at point d in Fig. 7.4.3 and the corresponding conditions are:

Relative humidity φ = 90 percent
Moisture content ω = 15 gm/kg
Reduction in moisture content   = 2 gm/kg
Reduction in specific enthalpy, point c to point $d = 10 \cdot 5$ kJ/kg.

The corresponding reduction in enthalpy of the airstream, ignoring the enthalpy of the condensed water, is:

$$0{\cdot}146 \times 10{\cdot}5 = 1{\cdot}53\ kW$$

7.4.2 Discussion of Results

It will be evident that the psychrometric chart provides an extremely clear and simple method of presenting air conditioning processes. A great deal of information can be deduced from two simple measurements: the wet- and dry-bulb temperatures.

It will be noticed that the cooling process, state c to state d, although it results in the deposition of moisture from the air, does not, as might be expected, end on the saturation line but at a relative humidity of 90 percent. This is because not all the air passing the cooler is brought into sufficiently close contact with the cooling surfaces to ensure that it is cooled to the saturation temperature.

The measured rate of water deposition in the cooling process agrees closely (to within less than 2 percent) with the amount calculated from the wet- and dry-bulb temperatures. The measurement of the relative humidity by wet- and dry-bulb indications becomes progressively less exact as φ approaches 100 percent. This is clear from the psychrometric chart which shows, for instance, that at an air temperature of 30°C a change in relative humidity from 10 percent to 20 percent corresponds to a change in wet-bulb temperatures of 2·4°C, while a change from $\varphi = 90$ percent to $\varphi = 100$ percent corresponds to an interval in wet-bulb temperature of only 1·3°C.

7.4.3 Further Experiments and Questions

(a) Investigate the dependence of the wet- and dry-bulb temperatures on air velocity. What minimum velocity is to be recommended?

(b) Experiments aimed at establishing some desired final conditions of temperature and relative humidity involving prediction of the necessary processes from the psychrometric chart.

8

Combustion

8.1 Combustion Processes

The majority of the world's requirement for electrical power and space heating, representing perhaps two-thirds of total world energy consumption, is still met by the combustion of coal, gas or oil in various forms of combustion chamber, mostly operating at atmospheric pressure. Combustion of this kind may be considered as a steady-flow process, Fig. 8.1.1. It will be evident that in the process mass is conserved:

$$\dot{m}_p = \dot{m}_a + \dot{m}_f \tag{8.1}$$

where the suffixes a, f and p refer respectively to the air and fuel taking part in the process and to the products of combustion.

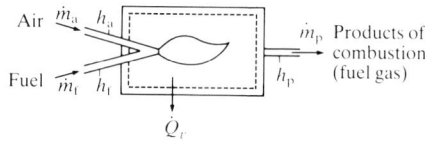

Fig. 8.1.1   Combustion: a steady-flow process

No mechanical work is performed in the course of the combustion process, and we can apply the First Law in a manner analogous to equation (4.5):

$$h_f + \frac{\dot{m}_a}{\dot{m}_f} \cdot h_a - \frac{\dot{m}_p}{\dot{m}_f} \cdot h_p - \Delta h = 0 \tag{8.2}$$

Here h represents the enthalpies of the fuel, air and products of combustion, while Δh is the enthalpy of combustion. This is the steady-flow energy equation applied to a combustion process. In the applications of this equation encountered so far the nature of the fluid or fluids concerned remains unchanged in the course of the process; in the present case the properties of the fluid at the end of the process, the products of combustion, or "flue gas", differ fundamentally from those present at the start.

The choice of temperature from which the enthalpy of the fluids is calculated affects the value of Δh that is deduced, thus differing from previous cases where we were only concerned with differences in enthalpy of the same substance. A solution that appears attractive in principle is to calculate enthalpies from the absolute zero temperature, but this presents serious practical difficulties. It is accepted practice to take 0°C as corresponding to zero enthalpy but to make experimental determinations of Δh at a temperature of 25°C. This implies that the air and fuel should be at a temperature of 25°C before combustion and that the flue gas should be cooled to this temperature after completion of the process. Values of the enthalpy of combustion determined experimentally at a temperature of 25°C are reported in two different ways:

Gross or Higher calorific value, Q_{gr}
Net or Lower calorific value, Q_{net}

The gross calorific value takes into account the latent heat of the water vapour produced by the combustion of the hydrogen that forms a constituent of all fuels, and when measuring the gross calorific value it must be ensured that this water vapour is condensed at the end of the process. The net calorific value does not include this latent heat, and it is assumed that the water produced remains in the form of vapour at the end of the process.

In practical combustion processes the water vapour is not condensed and the net calorific value forms the usual basis for calculating thermal efficiencies. Calorific values are defined in some detail in [1].* Table 8.1.1 shows gross and net calorific values for various fuel gases of technical importance. The presence of excess air in the combustion process has no effect on the measured calorific value as the enthalpy of the air not involved in combustion remains unchanged by the process.

Table 8.1.1   Calorific Value of Fuel Gases

	Gross calorific value kJ/kg	Net calorific value kJ/kg	Air/fuel ratio (stoichiometric)
Hydrogen, H_2	141 810	119 940	34·3
Carbon monoxide, CO	10 100		2·46
Methane, CH_4	55 520	50 020	17·24
Ethane, C_2H_6	51 880	47 490	16·10
Acetylene, C_2H_2	49 920	48 230	13·27
Ethylene, C_2H_4	50 300	47 160	14·78
Propane, C_3H_8	50 350	46 360	15·64
n-Butane, n-C_4H_{10}	49 730	45 730	15·43

The calorific value of solid and liquid fuels is determined experimentally in a bomb calorimeter, Fig. 8.1.2. This consists of a gas-tight steel vessel immersed in water. A weighed quantity of fuel is introduced into the calorimeter which is then charged with oxygen at a pressure in the region of 25 atmospheres and the fuel ignited electrically. The initial and final steady temperatures of the water in which the bomb is immersed

---

* See also BS 526:1961, Definitions of the Calorific Value of Fuels.

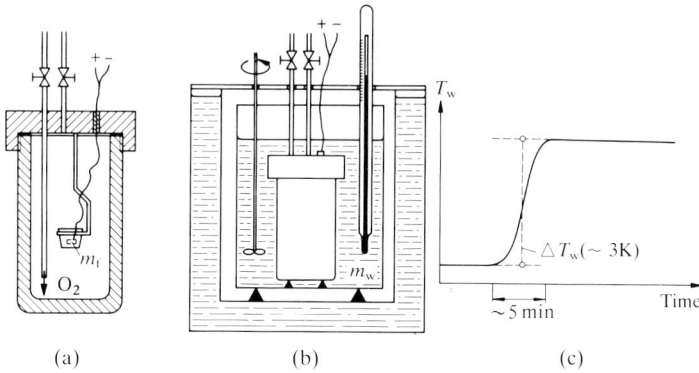

Fig. 8.1.2 (a) Bomb calorimeter (b) Calorimeter in water bath (c) Temperature change in bath

are noted and the temperature rise is limited to about 3°C. A separate experiment, using either electrical heating or a fuel of known calorific value, determines the water equivalent of the calorimeter. The use of oxygen rather than air for the atmosphere in which combustion takes place does not affect the calorific value determined by the apparatus.

Enough water is introduced into the bomb before combustion to ensure that the oxygen is saturated and the moisture produced from combustion of the hydrogen content of the fuel is therefore condensed: the calorimeter determines the gross calorific value.

For determining the calorific value of gaseous fuels combustion takes place continuously at atmospheric pressure in a water-cooled combustion chamber, and in this case also the gross calorific value is determined since the water vapour in the gas leaving the combustion chamber is condensed. If the hydrogen content of the fuel and either the gross or net calorific values are known, the other value may be calculated using known values of the enthalpy of water vapour.

8.2 Combustion Equations

The principal active constituents of all fuels of commercial importance are carbon and hydrogen, the only other element that contributes appreciable energy to the combustion process being sulphur, present as an impurity particularly in heavy fuel oils. Fuels may also contain combined oxygen (particularly in the case of alcohols) and ash. Some typical fuel compositions by weight are given in Table 8.2.1.

Table 8.2.1 Typical Fuel Compositions (by Mass)

	C	H	O	N + S	Ash
Anthracite	90	3	2	1	4
Diesel fuel	86	13		1	
Methanol	37·5	12·5	50		
Methane	75	25			
North Sea gas	72	24	0·5	3·5	

153

These elements are present in the fuel in a vast number of different chemical combinations, many of them unknown in the case of any particular fuel. Combustion takes place at high temperatures at which the chemical bonds between the different constituents of the fuel are broken, and we are concerned only with the reaction of the individual elements with oxygen:

Carbon, $C + O_2 = CO_2$
in mols: (1 mol = molecular weight in kg of the element or compound):

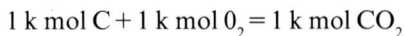

$$1 \text{ k mol } C + 1 \text{ k mol } 0_2 = 1 \text{ k mol } CO_2$$

in masses:

$$12 \text{ kg } C + 32 \text{ kg } O_2 = 44 \text{ kg } CO_2$$

Carbon, relative to 1 kg of fuel:

$$1 \text{ kg } C + 2 \cdot 664 \text{ kg } O_2 = 3 \cdot 664 \text{ kg } CO_2$$

Hydrogen, $2H_2 + O_2 = 2H_2O$

$$1 \text{ kg } H_2 + 7 \cdot 937 \text{ kg } O_2 = 8 \cdot 937 \text{ kg } H_2O$$

Sulphur, $S + O_2 = SO_2$

$$1 \text{ kg } S + 0 \cdot 998 \text{ kg } O_2 = 1 \cdot 998 \text{ kg } SO_2$$

In the case of pure gaseous fuels the exact chemical composition is usually known. As an example:

Propane, $C_3H_8 + 5H_2O = 3CO_2 + 4H_2O$

$$1 \text{ kg } C_3H_8 + 3 \cdot 628 \text{ kg } O_2 = 2 \cdot 994 \text{ kg } CO_2 + 1 \cdot 634 \text{ kg } H_2O$$

All these equations assume complete combustion. Sufficient oxygen or air must be available, otherwise incomplete combustion occurs resulting in the presence of H_2, CO or soot in the combustion products. A stoichiometric or "correct" mixture of air and fuel is one in which there is just enough oxygen present for the complete combustion of the fuel. The stoichiometric air/fuel ratio for a typical liquid internal combustion engine fuel is about 14·5:1, while the stoichiometric value for solid fuels is slightly lower than this. It is almost always necessary to provide a certain amount of excess air in order to ensure that combustion is completed during the usually extremely brief period available for the process. The efficiency of combustion in furnaces is very sensitive to the amount of excess air provided.

The excess air ratio e is defined as:

$$e = \frac{\text{actual air/fuel ratio}}{\text{stoichiometric air/fuel ratio}}$$

$e = 1$ corresponds to a "correct" mixture, while values of less than 1 indicate that insufficient air is present for complete combustion.

8.3 The Combustion Process

The ignition of a mixture of fuel and air may be regarded as the initiation of a chain reaction. Consider, for example, a mixture of methane and air, Fig. 8.3.1. Ignition of the mixture is brought about, for example by an electric spark, by imparting sufficient energy to a certain number of oxygen molecules to enable these to react with methane molecules. This reaction results in the production of CO_2 and H_2O molecules which by virtue of the energy released in the course of the combustion reaction are travelling at very high velocity. Almost immediately each newly formed molecule will strike either an N_2, a CH_4 or an O_2 molecule, the greatest probability being an impact with an N_2 molecule. After several impacts the energy of the CO_2 or N_2O molecule will have been dissipated and only occasionally will it give rise to a further reaction. (It is, of course, well known that in combustible mixtures, even at room temperature, occasional individual molecular reactions take place, set off, for example, by the random impact of energetic particles. A stoichiometric mixture of hydrogen and oxygen exposed to sunlight reacts in this way.)

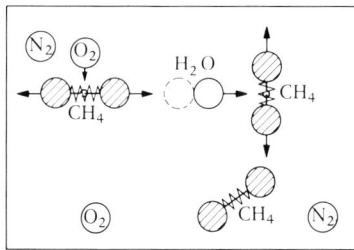

Fig. 8.3.1 Ignition as a chain reaction

If the mixture of fuel and air exceeds a certain critical minimum volume and the initial number of individual reactions is sufficiently great, the kinetic energy of the molecules constituting the products of combustion will not be dissipated as described above; instead a chain reaction will be initiated and will proceed through the volume of mixture in the form of a flame front.

From the above description of the combustion process in a mixture of fuel in gaseous or vapour form and air it may be suspected that the range of air/fuel ratios within which combustion can be initiated will be limited, and experience shows that this is the case. For example, in a petrol engine the range of air/fuel ratios within which the engine can operate lies within a "rich" limit of about 8:1 by weight and a "weak" limit of about 20:1. If the temperature of the mixture is sufficiently high, however, combustion can take place outside the range of air/fuel ratios applicable at atmospheric temperature, since the general level of kinetic energy of the molecules may be sufficient to initiate the reaction.

The velocity at which combustion takes place (the rate of travel of the flame front through the mixture) tends to be a maximum in the region of the stoichiometric mixture, as the probability of an energetic molecule striking an unburned molecule of fuel is then at its highest. Fig. 8.3.2 shows curves of velocity of propagation of the flame front against air/fuel ratio for several commercial gaseous fuels. These results may be obtained very simply, by passing a mixture of air and fuel gas through a transparent plastic tube of about 20 mm bore, igniting the mixture, and observing the

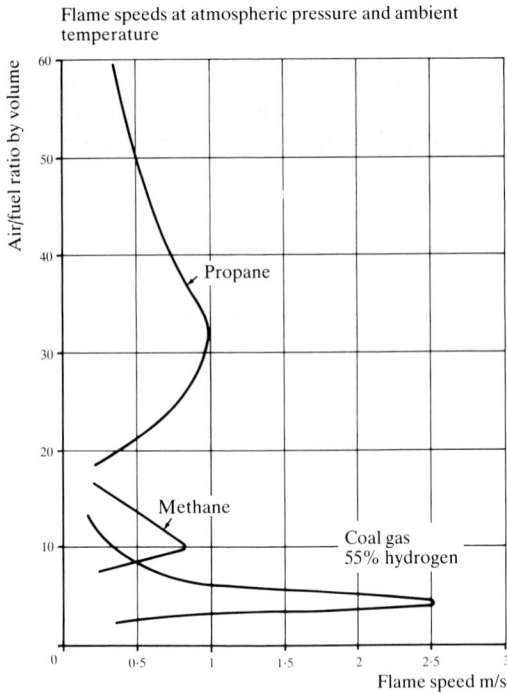

Flame speeds at atmospheric pressure and ambient temperature

Fig. 8.3.2 Relation between rate of propagation of flame front and air/fuel ratio for several commercial gaseous fuels

rate of flame travel. It will be observed that the maximum rate of flame propagation occurs in the neighbourhood of the stoichiometric air/fuel ratio.

In a conventional laboratory Bunsen burner the flame front assumes a conical form, the included angle at the apex of the cone adjusting automatically to ensure that the rate of flow of mixture through the flame front equals the rate of flame propagation. If a combustible mixture is passed through a wire gauze and then ignited, a stable flame front may persist immediately beyond the gauze. The velocity of flow is increased by the presence of the gauze and locally exceeds the velocity of flame propagation, inhibiting the travel of the flame front in an upstream direction.

The velocity with which the flame front advances is enormously influenced by the level of turbulence in the mixture; the operation of a high-speed internal combustion engine would be impossible in the absence of intense turbulence in the combustion chamber.

8.4   The Enthalpy–Temperature Diagram for Combustion Processes

Continuous combustion processes are conveniently represented on an enthalpy–temperature diagram, Fig. 8.4.1. The diagram shows two curves, one representing the variation of the enthalpy of the flue gas with temperature, and the other showing the enthalpy of the air and fuel prior to combustion. The curves are drawn for 1 kg of fuel and, in the simplest form of the diagram, for a stoichiometric mixture. Zero enthalpy is taken arbitrarily at 0°C and the interval between the curves is made equal to Δh, the

156

enthalpy of combustion, equation (8.2). For most practical purposes the enthalpy of combustion may be taken as equal to either the gross or net calorific value, the choice depending on the purpose for which the diagram is to be used. The errors consequent on the measurement of calorific values at initial and final temperatures not exactly equal and also greater than 0°C are quite negligible for engineering calculations.

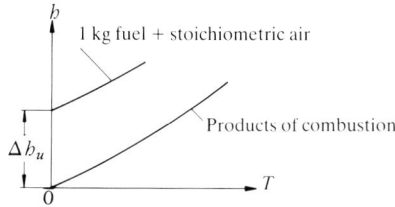

Fig. 8.4.1 Enthalpy–temperature diagram for combustion

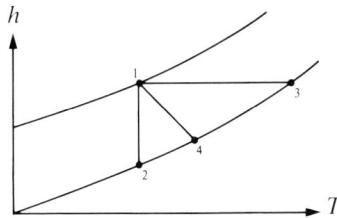

Fig. 8.4.2 Representation of various combustion processes on the enthalpy–temperature diagram

Both curves are related to temperature by an equation of the form:

$$h = c_{pm} \cdot t$$

where c_{pm} is the mean value of the specific heat at constant pressure between 0°C and the temperature t. c_{pm} is temperature dependent, increasing with temperature so that the curves are concave upwards. It is almost independent of pressure, although in any case most combustion processes for which analysis in this form is appropriate take place at or near atmospheric pressure.

Several combustion processes are illustrated in Fig. 8.4.2. 1–2 represents steady-flow combustion as in a gas calorimeter, in which the heat transferred is equal to the enthalpy of combustion and the initial and final temperatures are identical. This process results in the maximum possible transfer of energy to the surroundings. 1–3 represents steady-flow adiabatic combustion in which the enthalpy remains constant and there is merely a redistribution of energy resulting in a rise of temperature: chemical energy is transformed into sensible energy. This process results in maximum possible temperature rise. Most real combustion processes, such as that examined in Experiment 9, are accompanied by some heat transfer to the combustion chamber and may be represented on the diagram by lines such as 1–4.

It is usual to add parameter lines to the diagram for different values of excess air ratio e. Enthalpy–temperature diagrams can be constructed for any fuel, and Fig. 8.4.3 shows such a diagram for propane. With the exception of alcohols, which contain a considerable proportion of combined oxygen, and fuels such as low-grade

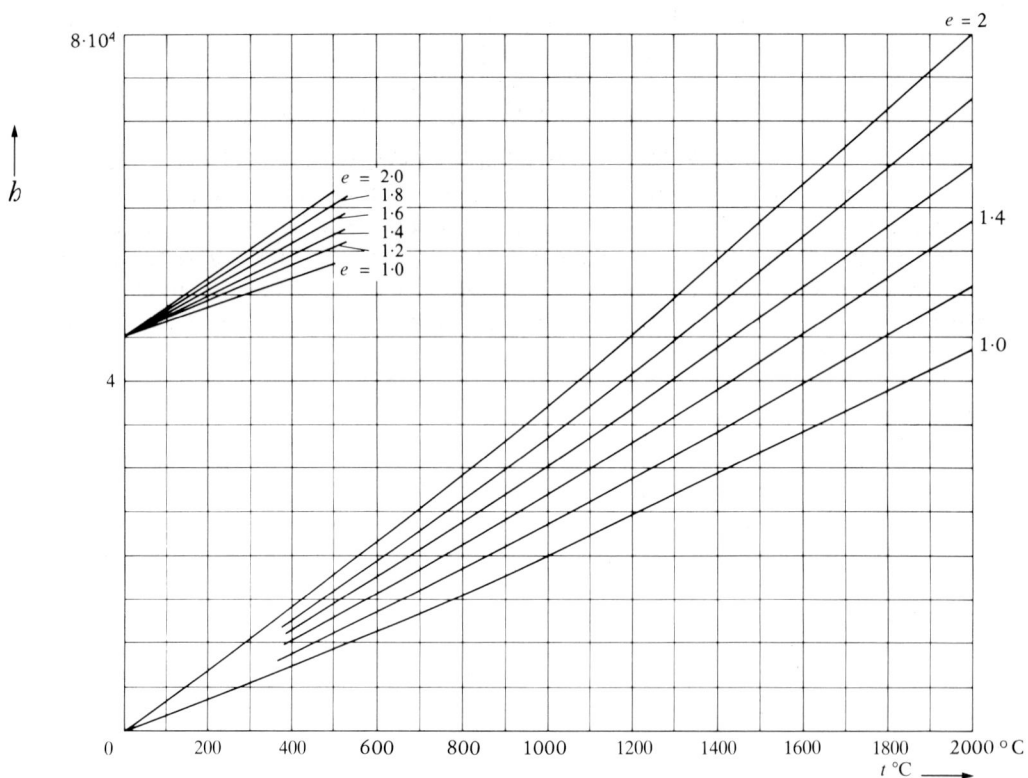

Fig. 8.4.3 Enthalpy–temperature diagram for combustion of propane, C_3H_8

coals or peat, with a high moisture content, most fuels when burnt stoichiometrically in air at atmospheric temperature produce adiabatic combustion temperatures within a fairly narrow range; roughly 2300–2700°K. This is understandable if one considers the roles of fuel and combustion air as reversed and regards the oxygen in the air as equivalent to fuel. The calorific values of a wide range of fuels and also of pure hydrogen and carbon lie within a comparatively narrow band when referred to a common mass of combustion air. In the case of the internal combustion engine there is no difficulty in introducing any desired quantity of fuel into the cylinder; the possible power output is limited entirely by the mass of the air charge, and it is important to ensure the largest possible energy release per unit mass of air.

The actual temperatures achieved in combustion are less than those that would be predicted from a knowledge of the enthalpy of combustion and the mean specific heats of the combustion products because of the phenomenon of dissociation. Above a temperature of about 1500°C the principal combustion reactions become to a significant extent reversible. In particular, CO_2 tends to dissociate into carbon monoxide and oxygen, water either into hydrogen and oxygen or into hydrogen and a hydroxyl radical, while nitrogen ceases to behave as an inert gas and combines with oxygen to form nitric oxide. All these reactions take place with the absorption of heat. They are represented by equations of the following form, in which the symbol for equality signifies that they may proceed in either direction:

158

$$CO_2 \rightleftarrows CO + \tfrac{1}{2}O_2$$
$$H_2O \rightleftarrows H + \tfrac{1}{2}O_2$$
$$H_2O_2 \rightleftarrows OH + \tfrac{1}{2}O_2$$
$$N_2 + O_2 \rightleftarrows 2NO$$

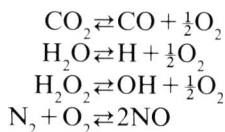

A theoretical treatment of the phenomenon is outside the scope of this book. It is possible to determine equilibrium or dissociation constants which indicate, for any temperature, the equilibrium state of these reactions. The degree of dissociation increases with temperature, as does the extent to which the heat released falls short of that predicted by calorimetric determinations of calorific value. As the temperature of the combustion products falls, recombination takes place and the sensible heat absorbed in the dissociation process is again released.

8.5 Flue Gas Analysis

Accurate control of combustion in power station boilers, industrial processes and industrial and domestic heating plants is essential on the grounds both of efficiency and for the minimizing of pollution. Unsatisfactory combustion conditions result in the presence of unburned carbon (soot or smoke), carbon monoxide and, under certain high-temperature conditions, of nitric oxides (NO_x) in the flue gas. Pollution is commonly associated with a deficiency in the air supply; on the other hand, the presence of excess air results in a decline in the efficiency of the process, since an unnecessarily high proportion of the enthalpy of combustion is carried away in the flue gas.

Accurate monitoring and control of flue gas composition is thus a matter of great practical importance. The traditional means of analysing flue gas was the Orsat apparatus, in which the proportions of CO_2, O_2 and CO in the flue gas are measured by successive absorption by appropriate reagents. The Orsat apparatus is slow and inconvenient in use, and furnace installations of any size are now usually equipped with permanent monitoring devices, of which the most sensitive is the gas chromatograph, which depends for its operation on the sensing of the small changes in thermal conductivity of flue gases associated with variations in composition.

A major problem, consideration of which is outside the scope of this book, concerns the production of SO_2 as a component of the flue gases from power stations and industrial furnaces. Sulphur dioxide produced in this way, together with oxides of nitrogen, is responsible for the phenomenon of "acid rain", which can give rise to grave ecological problems.

8.6 Experiment 9: Flow and Combustion Processes in a Gas Combustion Chamber

The combustion of fuels at or near atmospheric pressure, with heat transfer to the surroundings of the combustion chamber, forms a class of processes of great practical importance. The present experiment illustrates all the main features and can be performed in an apparatus of the type shown diagrammatically in Fig. 8.6.1.

The apparatus consists essentially of a water-jacketed combustion chamber which is supplied with air by a centrifugal blower and with fuel—in the case of the present

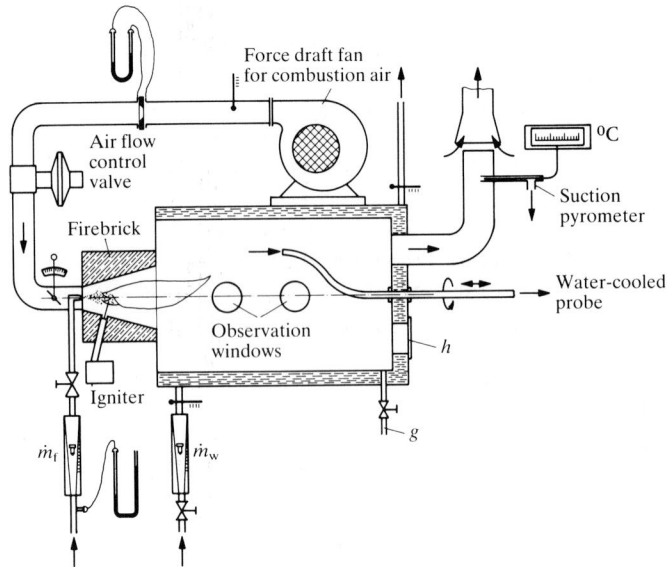

Fig. 8.6.1 Schematic arrangement of experimental combustion chamber

experiment, commercial propane. The mixture is initially ignited electrically and the first stage of combustion takes place within a conical chamber formed in a block of refractory material which, after a period of operation, becomes incandescent. Combustion continues in the main chamber which is surrounded by the water jacket, and the products of combustion leave the chamber by way of a flue, which is equipped with a suction pyrometer for measuring the flue gas temperature.

The combustion chamber is fitted with observation windows and instrumentation includes methods of measuring air, fuel and cooling-water flow quantities and temperatures. Samples of combustion gas may be withdrawn from various points in the combustion chamber.

It is necessary to run the apparatus for some time, perhaps 30 minutes, to achieve stable temperature conditions, and the approach of stability may be assessed by taking readings every 5 minutes until there is no significant change between successive observations. The principal variable of interest is the excess air ratio e. It will be found that combustion proceeds with reasonable stability over a range of e from about 0·7 to 2. The stoichiometric air/fuel ratio for propane, the fuel used in the authors' experiment, is 15·7:1 by weight (24:1 by volume).

It is of interest to observe the appearance of the flame through the window at the end of the combustion chamber and to note the changes in its appearance as e is varied from a value corresponding to insufficient air for complete combustion to one corresponding to a large excess of air.

As a preliminary experiment it is necessary to calibrate the suction pyrometer employed to measure the flue gas temperature. This instrument has been described in Section 2.8.2 and a calibration curve is shown in Fig. 8.6.2, which shows the relation between the temperature recorded by the thermocouple and the suction pressure. It will be observed that the reading rises from about 640° to about 700°C as increasing

160

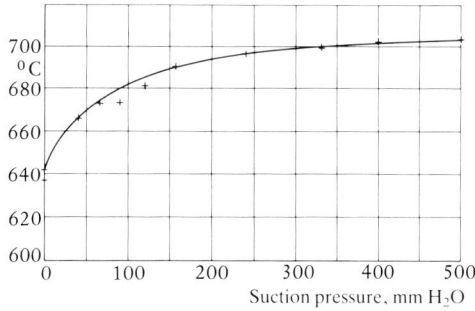

Fig. 8.6.2 Calibration curve for suction pyrometer

suction is applied to the pyrometer and the rate at which gas is drawn past the thermoelement increases. With depressions greater than 500 mmH$_2$O there is little change in recorded temperature, and it may be assumed that true readings of the flue gas temperature are given by this particular instrument when the depression exceeds this value.

8.6.1   Measurements and Calculations

The purpose of the experiment is to compare the heat released in the course of combustion with the theoretical heat of reaction (net calorific value Q_{net}) for propane. Fig. 8.6.3 shows a control volume with the relevant mass and enthalpy flows, while

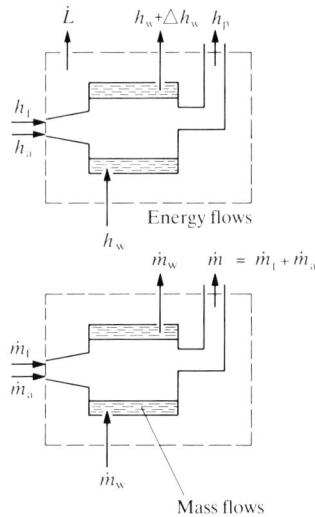

Fig. 8.6.3   Combustion chamber: control volume and energy flows

the process is represented schematically by a line such as 1–4 on the enthalpy–temperature diagram of Fig. 8.4.2. The steady-flow energy equation, referred to 1 kg of fuel, is written as follows:

$$h_f + \frac{\dot{m}_a}{\dot{m}_f} \cdot h_a - \frac{(\dot{m}_a + \dot{m}_f)}{\dot{m}_f} h_p - \dot{m}_w \Delta h_w - \dot{L} = 0 \tag{8.3}$$

161

where $\qquad h_f$ = enthalpy of propane at inlet temperature t_1

$\dot{m}_a/\dot{m}_f$ = air/fuel ratio

h_a = enthalpy of air at inlet temperature t_1

h_p = enthalpy of products of combustion at temperature t_2

$\dot{m}_w$ = rate of flow of cooling water

Δh_w = increase in enthalpy of cooling water

$\dot{L}$ = rate of heat loss from combustion chamber

The losses $\dot{L}$ include those due to convection and radiation from the combustion chamber plus any resulting from incomplete combustion of the fuel and any residual degree of dissociation still present in the flue gas.

For the present purpose equation (8.3) requires to be rewritten:

$$(Q_{net} + c_{pm,f} \cdot t_1) + \frac{\dot{m}_a}{\dot{m}_f} \cdot c_{pa} t_1 - \frac{\dot{m}_a + \dot{m}_f}{\dot{m}_f} h_p - 4 \cdot 1868(t_4 - t_3)\dot{m}_w - \dot{L} = 0$$

where

$c_{pm,f}$ = specific heat at constant pressure of propane vapour, mean value between $t = 0°C$ and $t = t_1$

c_{pa} = specific heat of air at constant pressure

t_3, t_4 = inlet, outlet temperatures of cooling water.

Q_{net} is taken from Table 8.1.1, while h_p is read from Fig. 8.4.3, which shows the enthalpy of propane flue gas for various values of air/fuel ratio e.

Table 8.6.1 shows one of a set of experimental observations, taken at a common rate of fuel flow of 6 kg/h (it is usual in combustion calculations to measure flow rates

Table 8.6.1 Experiment 9: Flow and Combustion Processes in a Gas Combustion Chamber: Observations

$\dot{m}_a$	kg/hr	100
$\dot{m}_f$	kg/hr	6·0
t_1	°C	46
t_2	°C	715
$\dot{m}_w$	kg/hr	900
t_3	°C	18·9
t_4	°C	61·2
ε		1·06
h_p	kJ/kg	14 500
$\dot{H}_p$	KJ/h	87 000
$\dot{H}_w$	kJ/h	159 100
$\dot{H}_a$	kJ/h	4600

per hour rather than per second). The tests are carried into the region of "rich" mixtures ($e < 1$) in which there is insufficient air for complete combustion. The curves of flue gas enthalpy, Fig. 8.4.3, do not extent into this region but as $c_{pm,f}$ does not change significantly we may make use of the curve for $e = 1$.

8.6.2   Discussion of Results

The results are plotted in Fig. 8.6.4 as curves of heat transfer to the cooling water and heat content of flue gas against the air/fuel ratio. The diagram also shows the accepted net calorific value of pure propane, C_3H_8, i.e. 46 360 kJ/kg, although it does not necessarily follow that the calorific value of the gas used in the experiment agrees exactly with this figure; commercial propane is a mixture of propane and other hydrocarbon gases having approximately the same pressure/temperature characteristics as the pure substance but not necessarily identical properties in other respects.

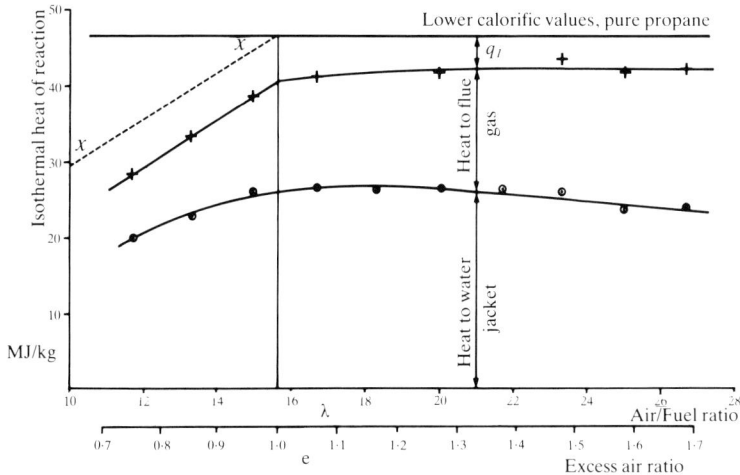

Fig. 8.6.4 Combustion of propane: relation between energy realized in combustion and excess air ratio

The dotted line xx shows the theoretical heat release rate under conditions of insufficient air supply on the assumption that the whole of the available oxygen is burned to CO_2 and H_2O and it will be observed that the experimental points lie closely parallel to this line. The curves make clear the undesirability of operating a combustion process with insufficient air supply; each 1 percent deficiency in air results in a reduction in the useful heat release of about 1·1 percent.

Operation with excess air also lowers the efficiency, though to a less marked extent. Thus an excess air ratio of 50 percent, corresponding to an air/fuel ratio of 23·6:1, results in a reduction in the heat transferred to the combustion-chamber water jacket of about 8 percent.

The difference between the net calorific value of pure propane and the observed heat release reaches a minimum at an air/fuel ratio of 21:1, corresponding to about one-third excess air, and then amounts to about 10 percent, an agreement that may be regarded as satisfactory. At the stoichiometric ratio the discrepancy has widened to 13 percent, probably indicating incomplete combustion.

8.6.3   Further Experimental Possibilities

(a) Observations of the appearance and stability of the flame for different air/fuel ratios.

163

(b) Analysis of CO_2 and CO content of the flue gas for different air/fuel ratios and comparisons with theory.

(c) An examination of the performance of thermocouple and other temperature-measuring instruments at various locations in the combustion chamber and flue under various combustion conditions.

(d) Measurements with other liquid and gaseous fuels.

(e) Examination of the effects of flame radiation on heat transfer using luminous and non-luminous flames.

9

Properties of Gases; Air as a Working Fluid

9.1 General Considerations

It is a remarkable fact that virtually all machines employed for the production of power from heat use one of two working fluids: steam or air. In this case "air" is loosely defined as including the products of combustion since internal combustion engines and gas turbines, with the exception of the closed-circuit gas turbine, operate on air for part of the working cycle and on combustion products for the remainder.

From time to time attempts have been made to introduce other working fluids, such as, for example, the early experiments on the use of mercury as the working fluid for a closed-circuit turbine, but these so far have shown little success. The universal availability and non-toxic nature of water and air present an overwhelming advantage. This situation is changing, particularly as a result of the need to devise practical thermodynamic cycles for the development of power from low-temperature sources such as solar collectors or from the small temperature differences available in the oceans. The thermodynamic properties of water are far from ideal, a major disadvantage being the extremely high pressures associated with quite moderate boiling temperatures.

The steam cycle has been discussed in Chapter 6 and the present chapter will be concerned with machines using air.

9.2 Properties of Gases

The First Law of Thermodynamics for unit mass of fluid undergoing a reversible process may be written:

$$q_{12} = u_2 - u_1 + W_{12} \tag{9.1}$$

where q_{12} = heat supplied to the fluid in the course of a change from state 1 to state 2, u_1 = initial internal energy, u_2 = final internal energy and W_{12} = work performed by the fluid on its surroundings.

If only displacement work is involved, we may write this equation in differential form:

$$dq = du + p \, dV \tag{9.2}$$

Enthalpy is defined as follows:

$$h = u + pV$$

In differential form:

$$dh = du + p\,dV + V\,dp \qquad (9.3)$$

Substituting in (9.2):

$$dq = dh - V\,dp \qquad (9.4)$$

These relationships are of a quite general nature and for fluids such as steam are not directly integrable. For the special case of a "perfect gas", however, various algebraic relationships may be shown to hold.

A perfect gas obeys the following equation of state:

$$pV = RT \qquad (9.5)$$

where R = gas constant, units J/kgK.

Real gases approximate closely to the equation at temperatures that are remote from the critical temperature but not so high as to give rise to dissociation (p. 158), and at moderate pressures. The agreement becomes closer as pressure is reduced.

It may be shown that for a gas that follows this equation u and h are necessarily independent of p and functions of T only. It may also be shown that c_p and c_v, the specific heats at constant pressure and constant volume, are either constant or functions of T only. The further assumption is made that c_p and c_v are in fact constant.

For a constant-volume process $dV = 0$ and equation (9.2) may be written:

$$dq = du = c_v\,dT$$

or, integrating:

$$u_2 - u_1 = c_v(T_2 - T_1) \qquad (9.6)$$

For a constant pressure process $dp = 0$ and equation (9.4) may be written:

$$dq = dh = c_p\,dT$$

or, integrating:

$$h_2 - h_1 = c_p(T_2 - T_1) \qquad (9.7)$$

From the definition of enthalpy, for a perfect gas:

$$h = u + RT \qquad (9.8)$$

Combining (9.6), (9.7) and (9.8):

$$c_p(T_2 - T_1) = c_v(T_2 - T_1) + R(T_2 - T_1)$$
$$c_p = c_v + R \qquad (9.9)$$

The ratio of the specific heats is given the symbol:

$$\frac{c_p}{c_v} = \gamma$$

We can now assemble the following simple relationships for a perfect gas:

$$dq = c_v\, dT + p\, dV$$
$$pV = RT$$
$$c_p = c_v + R$$

$$\frac{c_p}{c_v} = \gamma$$

In the case of an isothermal process:

$$dq = c_v \cdot 0 + p\, dV = RT\frac{dV}{V}$$

Integrating,

$$q_2 - q_1 = RT \log \left(\frac{V_2}{V_1}\right) = RT \log \left(\frac{p_1}{p_2}\right)$$
$$= p_1 V_1 \log \left(\frac{V_2}{V_1}\right)$$

and $\qquad p_1 V_1 = p_2 V_2$

In the case of an isentropic process:

$$0 = c_v\, dT + p\, dV$$
$$0 = c_v\, dT + \frac{RT}{V}\, dV$$
$$0 = c_v\frac{dT}{T} + \frac{R\, dV}{V}$$

Integrating,

$$c_v \log \left(\frac{T_1}{T_2}\right) = R \log \left(\frac{V_2}{V_1}\right)$$

$$\frac{T_1}{T_2} = \left(\frac{V_2}{V_1}\right)^{\frac{R}{c_v}} = \left(\frac{V_2}{V_1}\right)^{\gamma - 1}$$

A number of relationships between p, V ànd T for the isentropic expansion or compression of a perfect gas are summarized in Table 9.2.1.

9.3 Compression and Expansion Flow Processes for Air

The theoretical relationships derived in the previous section will be illustrated by a numerical example involving air as the gas concerned.

We consider first the two idealized processes that may be regarded as setting the standards with which real processes are compared: frictionless adiabatic and frictionless isothermal compression and expansion.

Fig. 9.3.1 represents the frictionless adiabatic compression of 1 kg of air from initial

Table 9.2.1 Expansion and Compression of a Perfect Gas with Constant Specific Heats: Isentropic Process

$$pV^{\gamma} = \text{constant} \qquad \frac{p_1}{p_2} = \left(\frac{V_2}{V_1}\right)^{\gamma} \qquad \frac{T_1}{T_2} = \left(\frac{V_2}{V_1}\right)^{\gamma-1} = \left(\frac{p_1}{p_2}\right)^{\frac{\gamma-1}{\gamma}}$$

Flow process

$$\int V\,dp = h_1 - h_2 = c_p(T_1 - T_2) = \frac{\gamma}{\gamma-1}p_2 V_2\left[\left(\frac{p_1}{p_2}\right)^{\frac{\gamma-1}{\gamma}} - 1\right]$$

Non-flow process

$$\int p\,dV = u_1 - u_2 = c_v(T_1 - T_2) = \frac{1}{\gamma-1}p_2 V_2\left[\left(\frac{p_1}{p_2}\right)^{\frac{\gamma-1}{\gamma}} - 1\right]$$

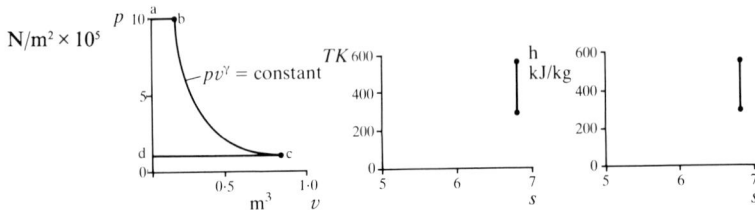

Fig. 9.3.1 Frictionless adiabatic compression

conditions $p = 10^5$ N/m², $T = 288$K, to a pressure $p = 10 \times 10^5$ N/m². There is no heat transfer to or from the air, and the process is thus represented on the T-s and h-s diagrams by a vertical line. The work W performed on the gas during the compression is represented by the area $abcd$ on the p-V diagram, and the steady-flow energy equation indicates that the increase in enthalpy equals the work performed. A frictionless adiabatic expansion may be represented by identical curves on the p-V, T-s and h-s diagrams.

A frictionless isothermal compression from the same initial conditions to the same final pressure is represented in Fig. 9.3.2 on the p-V, T-s and h-s diagrams. The area $abcd$ on the p-V diagram represents the compression work, and the area $abcd$ on the T-s diagram represents the equivalent heat transfer to the surroundings. The process is again reversible and an isothermal expansion is represented by identical curves.

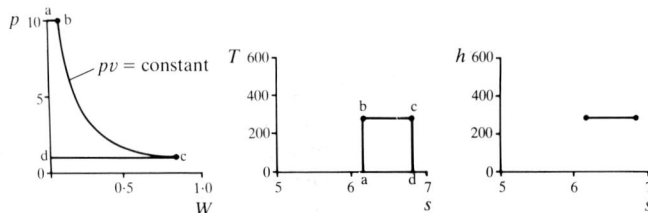

Fig. 9.3.2 Frictionless isothermal compression

An isothermal compression invariably requires the performance of less work upon the gas than an adiabatic between the same pressure limits, Fig. 9.3.3. Similarly, an adiabatic expansion results in the performance of less work than an isothermal.

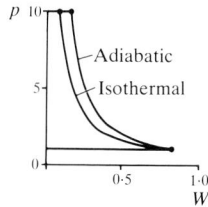

Fig. 9.3.3 *p-V* diagram for isothermal and adiabatic compression and expansion

A frictionless non-adiabatic compression process can follow any number of different paths. Fig. 9.3.4 shows such a process, defined by the relation $pV^{1.2} = \text{con}$ stant, in the course of which some heat has been transferred to the surroundings. On the *T-s* diagram the area *abcd* represents the heat transfer to the surroundings.

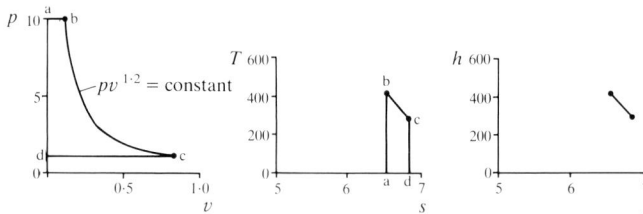

Fig. 9.3.4 Compression with heat transfer from gas

The s.f.e.e. indicates that the sum of the increase in enthalpy and the heat transferred to the surroundings is equivalent to the work done upon the gas. Once again we may represent the reverse process of frictionless non-adiabatic expansion accompanied by the reception of heat from the surroundings by identical curves.

Fig. 9.3.5 represents the case, important in practice, of frictionless non-adiabatic compression with the reception of heat from the surroundings. This particular process is represented by the polytropic $pV^{1.6} = \text{constant}$. In this case the area *abcd* on the *T-s* diagram represents the heat transferred to the gas from the surroundings.

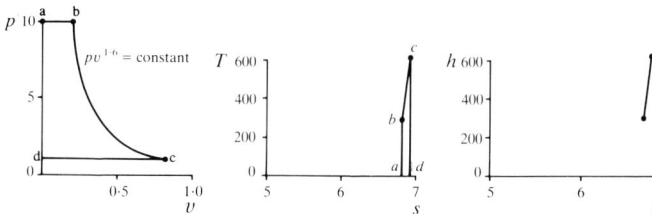

Fig. 9.3.5 Compression with heat transfer to gas

169

So far we have considered only reversible processes. All real gaseous expansion and compression processes involve a degree of irreversibility, and we consider first a process that occurs in all internal combustion and air expansion engines and which is completely irreversible: the process of partially unresisted expansion, such as takes place on the opening of the exhaust valve at the end of the expansion stroke. A volume of gas at an elevated pressure p_1 is suddenly released. We can say nothing about the intermediate states of the gas, but the First Law equation for the process may be written:

$$U_a - U_b = W$$

where W is the work performed by the gas, assuming no heat transfer with the surroundings. We may assess W as follows. The gas, originally occupying a volume V_1, has expanded to volume V_2 and in doing so has displaced a volume $(V_2 - V_1)$ of the surrounding atmosphere, performing displacement work $W = p_2(V_2 - V_1)$. This implies a reduction in internal energy and a fall in temperature.

We can plot the initial and final state points on the p-V and T-s diagrams, but we join them with a dotted line indicating that this does not represent identifiable intermediate states, Fig. 9.3.6.

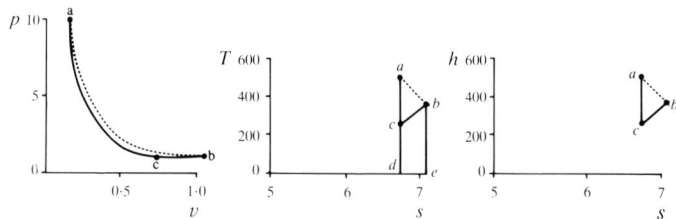

Fig. 9.3.6 Unresisted expansion

We can, however, visualize the process as follows. In the course of its escape from the containing vessel each successive element of gas is propelled by the gas behind it and also expands as its pressure falls. The gas thus emerges at high velocity but, meeting the surrounding atmosphere, the whole of this energy is rapidly dissipated in turbulence and reappears as sensible heat. We may thus regard the process as thermodynamically equivalent to frictionless adiabatic expansion ac, followed by the reception of heat at constant pressure, cb, the heat received being equal to the proportion of the work performed by the gas that is converted into kinetic energy. The process of expansion with friction, such as occurs in steam or gas turbines, may be represented in a similar way.

An analogous situation arises in the case of a centrifugal or axial-flow compressor. Part of the pressure rise is brought about by imparting a high velocity to the gas flowing through the machine and subsequently converting the associated kinetic energy into pressure energy in a diffuser, a process inherently prone to losses associated with turbulence.

A compression process in such a machine when represented on the p-V and T-s diagrams, Fig. 9.3.7, may appear to be identical with a frictionless compression process accompanied by heat reception from the surroundings, Fig. 9.3.5. It is,

170

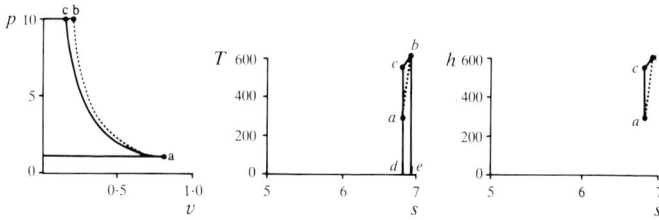

Fig. 9.3.7 Irreversible adiabatic compression with friction

however, essentially different in that the p-V diagram no longer represents work performed upon the gas, neither does the area beneath the T-s diagram represent heat transferred to it from the surroundings. We represent the process by dotted lines, and may then regard it as equivalent to a frictionless adiabatic compression ac to pressure p_2, followed by the addition of heat at constant pressure to reach the final state point b. This additional heat, which is represented by the area $bcde$ on the T-s diagram, represents a proportion of the work performed on the gas during the compression which is subsequently dissipated as heat as a result of friction and turbulence within the gas.

The First Law, in the form of the steady-flow energy equation, still applies to the process, and is written:

$$h_b - h_a = W_{IR} + W_{ac}$$

where W_{IR} is the irreversible work of compression, subsequently converted into heat, and W_{ac} is the frictionless adiabatic work corresponding to the compression process ac.

9.4 The Polytropic Process

In the last section we have represented certain processes by an equation of the form:

$$pV^n = \text{constant}$$

the polytropic equation, where n has values within the range:

$$n = 0: p \text{ constant}$$
$$n = 1: \text{isothermal}$$
$$n = \gamma: \text{adiabatic}$$
$$n \to \infty: V \text{ constant}$$

Processes involving heat transfer and/or friction are represented by intermediate values of n chosen to meet the specified end-states.

Note that there is no physical reason for assuming that an actual process linking end-states p_1, V_1 and p_2, V_2, and involving heat transfer and friction, should obey a law of the form $pV^n = \text{constant}$; this implies a specific relationship between the rates at which work is performed, heat is received or rejected, and losses occur, that may not correspond with the real situation.

However, the real process cannot greatly diverge from that implied by the polytropic relationship, which has the advantage of yielding specific algebraic

171

relationships between the quantities involved. These are summarized for a *reversible* polytropic process in Table 9.4.1. The expression for q_{12} represents heat transfer to (+) or from (−) the gas.

Table 9.4.1 Reversible Polytropic Process

$$pV^n = \text{constant} \quad \frac{p_1}{p_2} = \left(\frac{V_2}{V_1}\right)^n \quad \frac{T_1}{T_2} = \left(\frac{V_2}{V_1}\right)^{n-1} = \left(\frac{p_1}{p_2}\right)^{\frac{n-1}{n}}$$

$$\int V \, dp = \frac{n}{n-1} p_2 V_2 \left[\left(\frac{p_1}{p_2}\right)^{\frac{n-1}{n}} - 1 \right]$$

Flow process

$$\int p \, dV = \frac{1}{n-1} p_2 V_2 \left[\left(\frac{p_1}{p_2}\right)^{\frac{n-1}{n}} - 1 \right]$$

Non-flow process

$$q_{12} = \frac{\gamma - n}{(n-1)(\gamma-1)} p_2 V_2 \left[\left(\frac{p_1}{p_2}\right)^{\frac{n-1}{n}} - 1 \right]$$

Flow and non-flow processes

Expansion	$n < \gamma$	$q + \text{ve}$
Expansion	$n > \gamma$	$q - \text{ve}$
Compression	$n < \gamma$	$q - \text{ve}$
	$n > \gamma$	$q + \text{ve}$
	$n = \gamma$	$q = 0$

9.5   Experiment 5: Further Analysis

We are now in a position further to analyse the performance of the air motor, the subject of Experiment 5. A particular test yielded the following values of temperature and pressure:

$$p_1 = 4 \cdot 019 \times 10^5 \, \text{N/m}^2$$
$$T_1 = 301 \cdot 7 \text{K}$$
$$p_2 = 1 \cdot 019 \times 10^5 \, \text{N/m}^2$$
$$T_2 = 271 \cdot 1 \text{K}$$
$$V_1 = 0 \cdot 215 \, \text{m}^3/\text{kg}$$
$$V_2 = 0 \cdot 764 \, \text{m}^3/\text{kg}$$

From Table 9.4.1 we may deduce the corresponding polytropic index as:

$$n = 1 \cdot 084$$

If the process in the air motor were frictionless and reversible, Table 9.4.1 would indicate the following values for expander work and heat transfer to the air:

$$W_{12} = \int V \, dp = 112 \cdot 6 \text{ kJ/kg}$$
$$q_{12} = 82 \cdot 1 \text{ kJ/kg}$$

The actual mechanical output of the machine was only $W_{12} = 34 \cdot 1 \text{ kJ/kg}$, the decrease in enthalpy of the air in its passage through the machine 29·9 kJ/kg and, by difference, the heat transferred to the air in the course of the expansion, 4·2 kJ/kg.

The very large discrepancy between the actual power output and that corresponding to a frictionless polytropic is explained by losses, associated principally with non-expansive working, mechanical friction and internal leakage. We may imagine the process in the machine as taking place in two stages: a frictionless adiabatic followed by the dissipation of a proportion of the work performed by the gas during this expansion as heat. In the course of the adiabatic expansion, from Table 9.2.1 or Table 9.4.1:

$$W_{Ad} = 98 \cdot 1 \text{ kJ/kg}$$

We can write the following energy balance:

	kJ/kg
Adiabatic work	98·1
Heat received	4·2
	102·3
Work dissipated in friction and turbulence	68·2
Net work output	34·1
	102·3

The isentropic efficiency η_{Ad} of an expander is defined as:

$$\frac{\text{work output of machine}}{\text{work output of ideal adiabatic machine}}$$

In the present case:

$$\eta_{Ad} = \frac{34 \cdot 1}{98 \cdot 1} = 34 \cdot 8 \text{ percent}$$

A fairer basis of comparison for the performance of the air motor, since it operates non-expansively, would be the performance of an ideal non-expansive machine, for which

$$W_{12} = (p_1 - p_2)V_1 = 64 \cdot 5 \text{ kJ/kg}$$

giving an efficiency:

$$\eta = \frac{34 \cdot 1}{64 \cdot 5} = 52 \cdot 9 \text{ percent}$$

It is possible, by comprehensive tests over a range of speeds and supply pressures, to quantify the other sources of loss in the air motor and thus to account for the remaining shortfall in power output, amounting to $64 \cdot 5 - 34 \cdot 1 = 30 \cdot 4 \text{ kJ/kg}$.

It is found that losses due to leakage of air through the machine and mechanical losses are of roughly the same magnitude, while a further loss, associated with flow resistance, increases with motor speed but contributes only a small proportion of the total.

9.6 Air Compression

In the last Section the performance of the small air motor of Experiment 5 has been assessed by comparing its power output with that of two hypothetical ideal machines operating on the same air supply: a frictionless adiabatic machine and a frictionless machine operating on a non-expansive cycle.

It is of the greatest practical importance to establish standards by which the performance of real machines may be assessed, and in the case of air compressors, a very important class of industrial machine, two standards are available: the ideal adiabatic compressor and the ideal isothermal compressor. We have discussed these ideal processes above, and the power input required to compress a mass $\dot{m}$ kg/s of air at temperature t_1 from pressure p_1 to t_2 may be derived from the equations given in Table 9.2.1 and written:

Adiabatic:

$$P_{Ad} = \dot{m} \; \frac{\gamma}{\gamma - 1} \; RT_1 \left[\left(\frac{p_2}{p_1} \right)^{\frac{\gamma-1}{\gamma}} - 1 \right] \tag{9.10}$$

Isothermal:

$$P_{Is} = \dot{m} \; RT_1 \; \log_e \left(\frac{p_2}{p_1} \right) \tag{9.11}$$

The corresponding power inputs are represented on the pV diagram in Fig. 9.3.3. The isothermal compressor requires less power than the adiabatic machine, but it is not in all cases realistic to take this more efficient process as the standard by which actual machines should be judged. In the case of the high-speed centrifugal compressor that forms the subject of the next experiment there is no possibility of abstracting any appreciable amount of heat from the air in the course of compression and thus approximating to the isothermal process: the adiabatic efficiency is the appropriate measure of performance. The compression process in a low-speed multistage water-cooled piston compressor with inter-cooling on the other hand, represented diagrammatically in Fig. 9.6.1, may approximate to isothermal compression, and the isothermal efficiency is the appropriate basis of assessment.

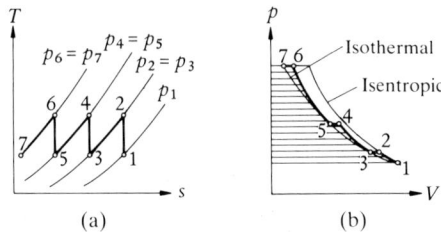

Fig. 9.6.1 Multi-stage adiabatic compression with intercooling:
(a) on *T-s* diagram   (b) on *p-V* diagram

The professional engineer is frequently called upon to make a decision as to the type of machine that it is appropriate to employ in a particular situation. In the case of air compressors there is a particularly wide range of choice, summarized in Fig. 9.6.2, which shows the ranges of flow quantity and delivery pressure for which each type of machine is best suited.

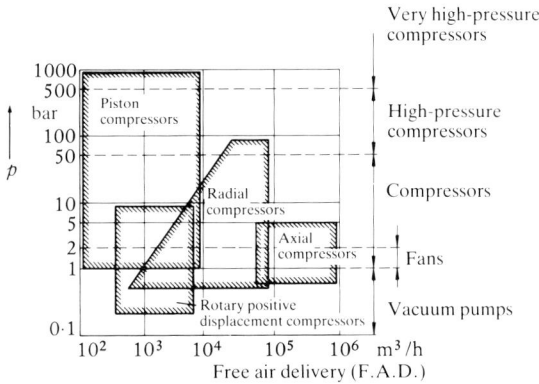

Fig. 9.6.2 Range of application of various types of compressor

9.7 Experiment 10: Performance of a High-speed Centrifugal Compressor

The centrifugal compressor occupies the middle ground between axial flow and positive displacement machines, Fig. 9.6.2. It is capable of larger pressure-ratios than the axial machines and suffers relatively less from leakage problems. Flow conditions are of extreme complexity and are as yet not fully understood. We may, however, distinguish two mechanisms by which the machine develops a pressure difference:

(a) By centrifugal forces acting upon the air rotating with the impeller. This pressure difference is present even with no throughput of air.

(b) By interaction between the flowing air and the rotor vanes and by the action of the diffuser surrounding the rotor, in which part of the kinetic energy imparted to the air in its passage through the rotor is transformed into pressure.

The test plant employed by the authors is shown in Fig. 9.7.1, a cross-section of the compressor in Fig. 9.7.2, while Fig. 9.7.3 is a schematic layout of the equipment. The impeller is driven by way of a variable ratio belt drive and an internal ball-bearing planetary step-up drive.

The great majority of small centrifugal compressors in service are employed for the supercharging of diesel engines, but they are then invariably driven by a gas turbine through which the engine exhaust is passed.

The authors' compressor draws air from the atmosphere by way of a calibrated nozzle and a throttling valve and discharges by way of a further throttling valve and a silencer. Suction and discharge pressures are indicated by mercury manometers, the pressure across the calibrated nozzle by a water manometer, and the various relevant temperatures by mercury-in-glass thermometers.

175

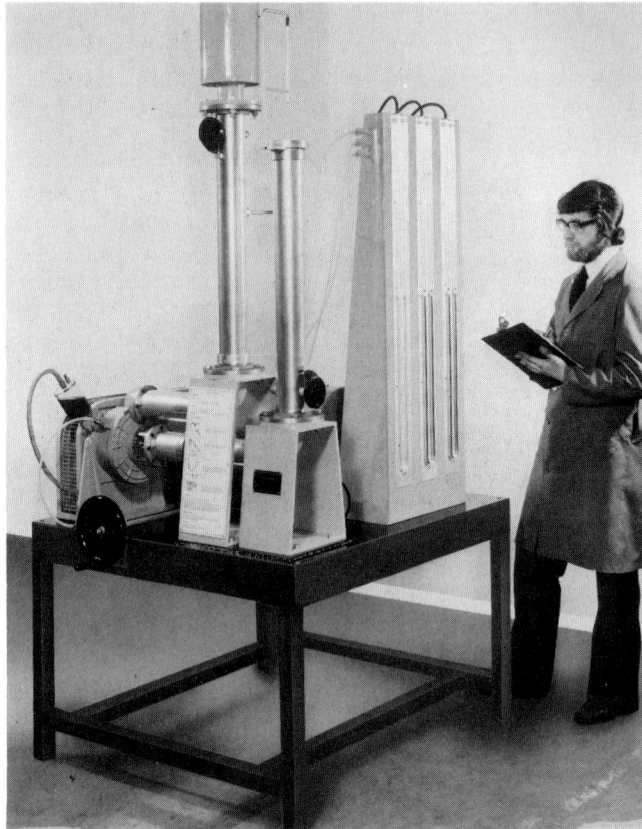

Fig. 9.7.1 Centrifugal compressor test rig

The main characteristics of the authors' machine were:

Speed range	22 000–29 000 rev/min.
Maximum pressure ratio	1·35:1
Impeller diameter	145 mm
Diameter of measuring nozzle	$d = 60$ mm
Coefficient of discharge	$C_d = 0·99$

Air, initially at rest at atmospheric conditions, expands through the measuring nozzle. The pressure difference across the nozzle, which may reach 700 mmH$_2$O, is such that compressibility may not be neglected. This is allowed for by the inclusion of a compressibility factor ε in the equation for $\dot{m}$ [23]:

$$\dot{m} = C_d \varepsilon \cdot \frac{\pi d^2}{4} \sqrt{2 \Delta p \rho_0} \qquad (9.12)$$

where Δp = pressure drop across measuring nozzle
   ρ_0 = density of air upstream of nozzle

The construction of the machine is such that it is not readily possible to measure

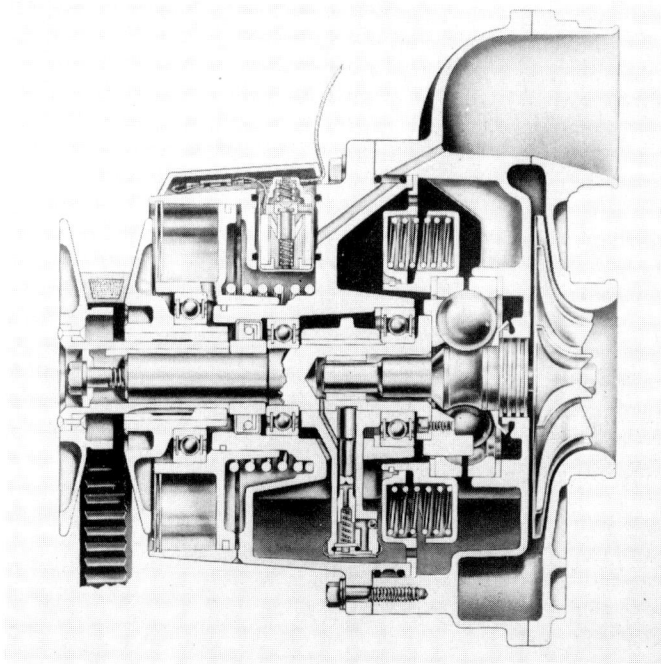

Fig. 9.7.2.   Transverse section of mechanically driven centrifugal compressor

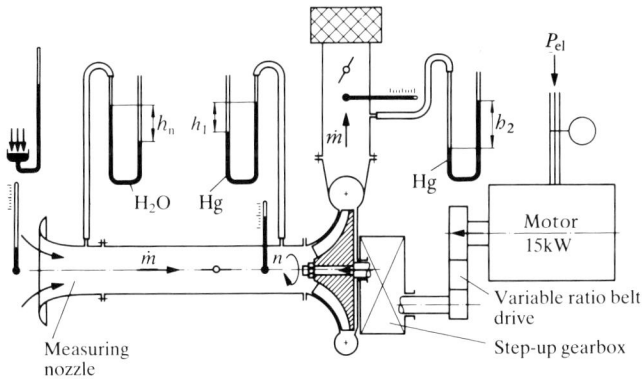

Fig. 9.7.3   Schematic arrangement of centrifugal compressor test rig

directly the power input to the compressor rotor and the following approximate method is employed. The electrical power input to the motor E_e is measured by a watt-meter and the corresponding mechanical power output E_m determined in a preliminary test. The relation between these quantities may be represented with sufficient accuracy by the equation:

$$E_m = 0.927(E_e - 1385) \qquad (9.13)$$

The unit is then run with the impeller of the compressor removed and the electrical power input E_e observed over a range of speeds. It may then be assumed that the corresponding motor power output E_m calculated from equation (9.13) equals the total friction losses E_f in the compressor drive. Results of such a test on the authors' machine are shown in Table 9.7.1. On the assumption that the mechanical losses in

Table 9.7.1   Relation between Compressor Drive Losses and Speed

n rev/min	22 000	23 000	24 000	25 000	26 000	27 000	28 000	29 000
Losses E_f W	1610	1690	1780	1860	1890	1890	1870	1860

the drive system at a given impeller speed are independent of the impeller power input E_s, we may write:

$$E_s = E_m - E_f \qquad (9.14)$$

This assumption is clearly not entirely justified; some increase in losses will occur with increase in the power transmitted through the drive system, leading to an overestimate of E_s and a consequent underestimate of the compressor efficiency. We must admit an uncertainty of at least 5 percent in the measurement of impeller power input. The practising engineer frequently finds himself in the position of being obliged to make this kind of approximate measurement.

This machine has the peculiarity that the step-up gearbox, in which the bulk of the transmission losses occur, is integral with the compressor casing. A consequence is that the compressor cannot be regarded as adiabatic; the bulk of the heat generated by friction in the gearbox is transmitted to the air passing through the compressor.

By closing the discharge throttle valve the flow may be reduced and at small flows the machine will eventually start to surge. The onset of surging is indicated by a pulsating noise from the compressor discharge. Surge, which is an important feature of all centrifugal compressors, is associated with a sharp fall in delivery pressure and a rapid rise in the temperature of the air discharged. The source of the noise is a resonance in the diffuser and discharge system, and this oscillation may also be propagated through the machine to affect the inlet system. The "surge line", Fig. 9.7.5, *below*, defines the limit of stable operation of the machine.

9.7.1   Measurements and Calculations

The performance is analysed on the basis of the control volume shown schematically in Fig. 9.7.4. The control surface could also be drawn to include the inlet nozzle and the inlet and discharge throttling valves, but this would introduce an unnecessary complication by taking into account the kinetic energy of the air entering and leaving the compressor. This is approximately equal at inlet and discharge since the ducts on both sides of the machine are of the same diameter, and thus approximately cancels out in the energy balance. A single observation is analysed below.

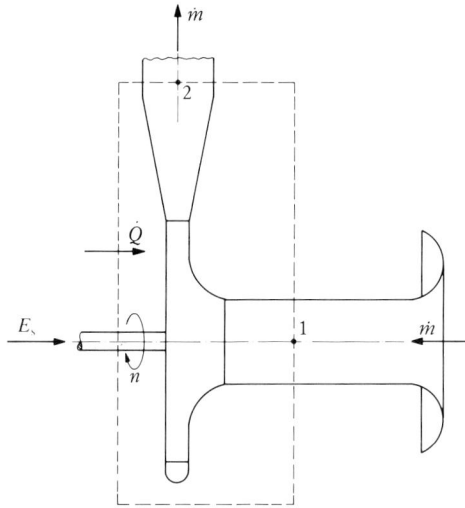

Fig. 9.7.4 Centrifugal compressor: control volume

Barometric pressure	$h_0 = 763$ mmHg
Compressor speed	$n = 26\,000$ rev/min.
Temperature at inlet	$t_0 = 23 \cdot 5°$C
Pressure difference across measuring nozzle	$h_d = 276$ mmH$_2$O
Pressure before compressor	$h_1 = -16$ mmHg
Temperature before compressor	$t_1 = 23 \cdot 5°$C
Pressure after compressor	$h_2 = +172$ mmHg
Temperature after compressor	$t_2 = 58 \cdot 5°$C
Electrical power input	$E_e = 10\,200$W
From equation (9.12)	$\dot{m} = 0 \cdot 222$ kg/s
($\varepsilon = 0 \cdot 986$, from [23])	

$$p_1 = 133 \cdot 3\,(763 - 16) = 99\,560\ \text{N/m}^2$$
$$p_2 = 133 \cdot 3\,(763 + 172) = 124\,600\ \text{N/m}^2$$
$$p_2/p_1 = 1 \cdot 252$$
$$E_m = 0 \cdot 927\,(10\,400 - 1385) = 8170\text{W} \quad \text{from equation (9.13)}$$
$$E_s = 8170 - 1890 = 6280\text{W} \quad \text{from equation (9.14)}$$

The s.f.e.e. for the complete machine (see Fig. 9.7.4) may be written:

$$E_s + \dot{Q} = \dot{m}(h_2 - h_1) = \dot{m}c_p(t_2 - t_1)$$
$$= 0 \cdot 222 \cdot 1004 \cdot (58 \cdot 5 - 23 \cdot 5)$$
$$= 7810\ \text{W}$$
$$\dot{Q} = 1530\ \text{W}$$

For an ideal adiabatic machine operating between the same pressure limits the required power input would be, Table 9.2.1:

$$E_{Ad} = \dot{m}\,\frac{\gamma}{\gamma - 1}\,RT_1\,\left[\left(\frac{p_2}{p_1} \right)^{\frac{\gamma - 1}{\gamma}} - 1 \right]$$
$$= 0 \cdot 222 \cdot 3 \cdot 5 \cdot 287 \cdot 296 \cdot 5\,(1 \cdot 252^{2/7} - 1)$$
$$= 4395\ \text{W}$$

179

The adiabatic efficiency of the machine:

$$\eta_{Ad} = \frac{E_{Ad}}{E_s} = \frac{4395}{6280} = 0\cdot70$$

The adiabatic efficiency based on the temperature rise of the air in its passage through the machine is given by:

$$\eta_T = \frac{4395}{7810} = 0\cdot563$$

Fig. 9.7.5 shows a curve of pressure ratio against flow rate at a speed of 26 000 rev/min. and also indicates the onset of surge at minimum flow rate and of choking at maximum flow. This latter phenomenon is associated with the onset of near-sonic velocities, usually in the inlet region of the impeller.

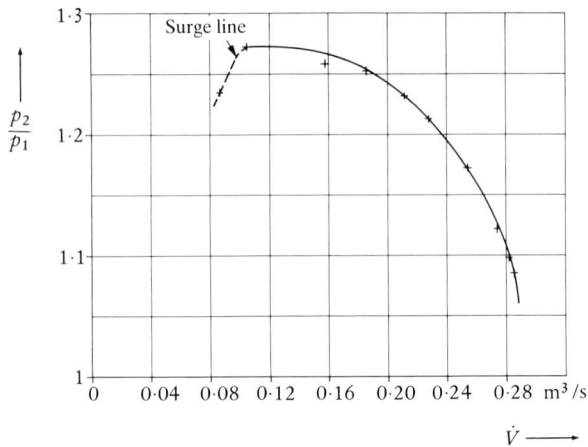

Fig. 9.7.5   Performance curve for centrifugal compressor at 26 000 rev/min

If curves such as that shown in Fig. 9.7.5 are plotted for a number of different speeds and the corresponding efficiencies calculated, the complete characteristic of the compressor may be constructed, Fig. 9.7.6.

9.7.2   Discussion of Results

This experiment is an excellent illustration of the amount of information regarding the performance of a machine that may be derived from the analysis of a comparatively small amount of correctly chosen experimental data. A peculiarity of the results is that the efficiency η_{Ad} calculated as the ratio of the power consumption of an ideal adiabatic machine to the shaft power input is substantially higher than the efficiency η_T based on the temperature rise through the machine. This is a direct consequence of the heat transfer, discussed above, from the compressor drive to the air passing through the machine.

9.7.3   Further Experiments

(a) Construction of the complete characteristic of the machine, Fig. 9.7.6.

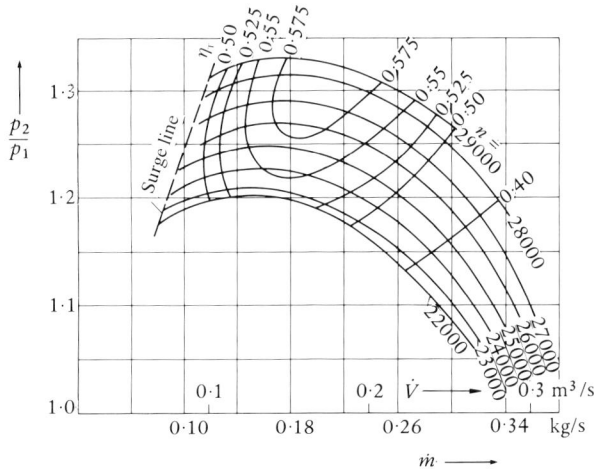

Fig. 9.7.6 Complete characteristic of centrifugal compressor

(b) Investigation of performance at reduced inlet pressures.

(c) Calculation of the dimensionless characteristics of the machine. The following dimensionless groups are commonly used to compare the performance of different machines:

$$\text{Pressure ratio} \qquad p_2/p_1$$

$$\text{Flow coefficient} \qquad \frac{\dot{m}\sqrt{c_p T_1}}{4p_1 r^2}$$

$$\text{Speed coefficient} \qquad \frac{2\omega r}{\sqrt{c_p T_1}}$$

Performance curves taken at different inlet pressures may be unified by plotting the results non-dimensionally.

(d) With additional instrumentation such as pressure tappings at strategic points in the machine casing, total head tubes in the diffuser, and pressure transducers capable of recording high-frequency fluctuations, studies may be made of:

> The compression process in the impeller;
> Pressure recovery in the diffuser;
> Flow patterns in inlet, impeller and diffuser;
> The phenomena of surge and choking.

9.8 The Gas Turbine

The classical open-cycle gas turbine is sketched in its simplest form in Fig. 9.8.1(a). It consists of a compressor, a combustion chamber, a turbine and a driven machine, typically an electrical generator. Fig. 9.8.1(b) shows the cycle on the enthalpy–entropy diagram drawn for a pressure ratio and maximum temperature typical of those encountered in a simple machine of moderate size:

181

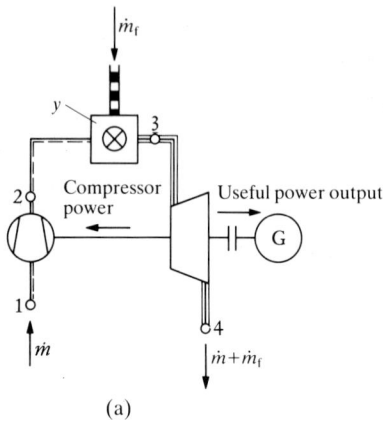

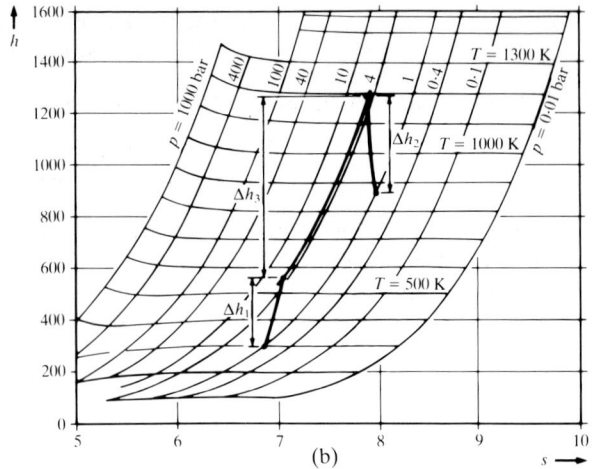

Fig. 9.8.1 (a) The simple open-cycle gas turbine (b) h-s diagram

Inlet pressure	$p_1 = 100\ kN/m^2$
Pressure ratio	$r = 5:1$
Combustion pressure	$p_2 = 500\ kN/m^2$
Adiabatic efficiency, compressor	0·85
Adiabatic efficiency, turbine	0·85
Air/fuel ratio	60:1

A common modification to this simple arrangement involves the splitting of the turbine into two units, a high-pressure unit which drives the compressor and a low-pressure unit which provides the useful power output. This is known as the two-shaft arrangement. Fuel economy may be improved by the provision of a heat exchanger in which the air delivered by the compressor is pre-heated by the turbine exhaust before reaching the combustion chamber.

It will be observed that the cycle is represented on the h-s diagram for air even though the working fluid after the combustion chamber consists not of air but of products of combustion. This, however, does not give rise to significant errors since even at maximum power a gas turbine operates with an excess air ratio of 300–400 percent. The mass of fuel represents only perhaps 2 percent of the mass of air. This limitation is necessary because of the limited maximum temperature, not exceeding perhaps 1000°C, to which it is permissible to subject the turbine blades. The permissible temperature is thus much lower than that associated with stoichiometric combustion. The relation between turbine power, compressor power, energy input and efficiency is given by:

Gross power output of turbine	$\dot{m}\ \Delta h_2$
Compressor input	$\dot{m}\ \Delta h_1$
Useful power output	$\dot{m}(\Delta h_2 - \Delta h_1)$
Heat input in combustion chamber	$\dot{Q} = \dot{m}\ \Delta h_3$
Efficiency	$\eta = \dfrac{\Delta h_2 - \Delta h_1}{\Delta h_3}$

182

For the simple machine for which the cycle is depicted in Fig. 9.8.1(b):

Thermal efficiency $= 0.225$
Work ratio $= 0.37$

If the adiabatic efficiency of compressor and turbine were each 100 percent,

Thermal efficiency $= 0.368$
Work ratio $= 0.53$

The overall thermal efficiency of the machine is clearly very sensitive to the component efficiencies.

The work ratio of a power cycle is defined, see p. 123, as the ratio of the net power output to the positive work. In the present case,

$$\text{Work ratio} = 1 - \frac{\Delta h_1}{\Delta h_2}$$

The work ratio of the simple gas turbine is rather unfavourable, one reason for the comparatively late development of the machine. Until it was possible to construct compressors and turbines having individual adiabatic efficiencies exceeding about 70 percent (after the year 1935) it was not possible to build a self-sustaining gas turbine as the power required to drive the compressor exceeded the power output of the turbine.

The thermal efficiency of the simple gas turbine is comparatively low, in the region of 20–25 percent, and the power output of single machines rarely exceeds 100MW, relatively small when compared with the steam turbine. The limitation in size is associated with permissible stress levels in the high-temperature components, with the avoidance of sonic velocities and with the comparatively limited working pressures in the open-cycle machine.

The work ratio of the simple gas turbine cycle may be improved by the adoption of multi-stage compression with inter-coolers between the stages, and by reheating the products of combustion in the course of the expansion process. Gas is withdrawn from the turbine, passed through a secondary combustion chamber in which further fuel is burned and returned to the turbine for further expansion. The effect of these modifications is to reduce the work of compression and increase the power output from the turbine, thus reducing the sensitivity of the overall performance to the component efficiencies.

A few closed-cycle gas turbines have been built. The combustion chamber is replaced by a heat exchanger in which the temperature of the gas in the closed circuit is increased and heat is removed from the gas between turbine outlet and compressor inlet in a further heat exchanger. The closed-circuit gas turbine has certain potential advantages over the open-cycle machine: gases other than air and having possibly more favourable thermodynamic properties may be used as the working fluid, while the circuit may be pressurized, permitting a higher power output from a machine of given size. The combination of a closed-circuit gas turbine with a nuclear reactor as the heat source presents certain attractive features.

The pure jet engine, Fig. 9.8.2, comprises an axial-flow compressor directly driven by an axial-flow turbine so proportioned that the power output of the turbine equals

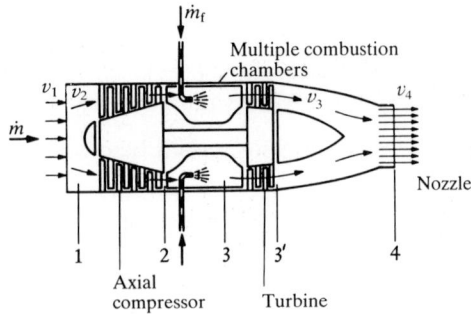

Fig. 9.8.2   The pure jet engine

the power input of the compressor (plus any additional power required for ancillary equipment).

The velocity v_1 of the air approaching the engine inlet is equal to the forward flight speed of the machine and considerably exceeds the relative velocity v_2 of the air entering the compressor. The static pressure at compressor inlet thus exceeds that of the surroundings: part of the compression process is achieved by decelerating the entering air (relative to the aircraft) on its way into the compressor. Similarly the pressure at turbine exit exceeds that of the surroundings, and the gas leaving the turbine at velocity v_3 is expanded in a nozzle to a velocity v_4 which exceeds the forward speed of the machine and thus gives rise to a thrust.

9.9   The Internal Combustion Engine

Like the gas turbine, the internal combustion engine operates on an open cycle in the course of which chemical as well as thermodynamic processes take place. While in the gas turbine compression, combustion and expansion take place in separate elements of the machine, in the internal combustion engine all three processes take place intermittently and in sequence in a single working space, the volume of which changes cyclically. The internal combustion engine was evolved from the steam engine by reasoning somewhat on the following lines: in the steam engine fuel is burned under a boiler in which heat is transferred to water giving rise to steam which, acting upon a piston, develops a surplus of power over that required to force the feed water into the boiler. Was there some means of shortening this sequence by burning the fuel directly in the cylinder?

Fig. 9.9.1 shows schematically the construction of the classical internal combustion engine. The working space is defined by a piston which reciprocates in a cylinder. At the uppermost position of the piston (top dead centre) the volume of the working space is at a minimum, the combustion chamber volume V_c. At the bottom position of the piston, bottom dead centre, the volume of the working space is at a maximum and the size of an internal combustion engine is frequently defined by the swept volume V_s, the difference in the volume of the working space between top and bottom dead centres. The compression ratio of the engine is defined as:

184

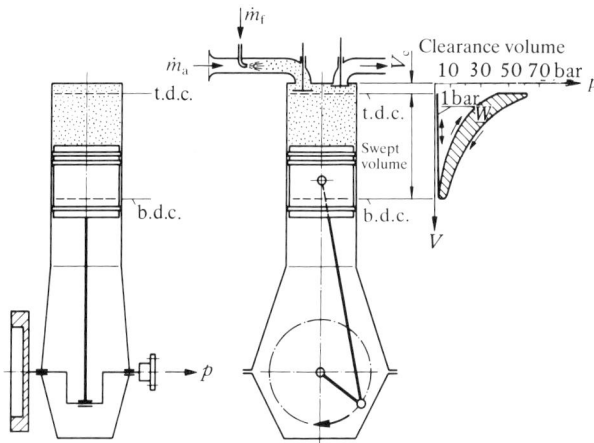

Fig. 9.9.1 Schematic arrangement of the simple internal combustion engine

$$r = \frac{V_s + V_c}{V_c}$$

The piston is reciprocated by means of an arrangement of crankshaft and connecting rod, and cyclic variation in the angular velocity of the crankshaft, consequent on the cyclic variation in the torque exerted by the piston, is reduced by fitting a massive flywheel. The flow of air to the cylinder and of exhaust gas from it is controlled by valves, or in some cases by ports in the cylinder wall, and fuel is introduced either into the airstream on its way to the engine in the conventional petrol engine or by injection directly into the working space in the diesel or direct-injection petrol engine.

Combustion is initiated either by a sparking plug or, in the case of the diesel engine, by direct transfer of heat from the heated and compressed air in the cylinder to the injected fuel. Fig. 9.9.2 shows the four successive phases of operation in an open combustion chamber diesel engine. In a four-stroke engine of this kind two complete revolutions of the crankshaft are required to complete the working cycle: one revolution in the course of which the exhaust gas is expelled and the cylinder charged with fresh air and one revolution in the course of which compression, combustion and expansion take place. In a two-stroke engine the complete process takes place in one revolution of the crankshaft, with the processes of exhaust and recharging of the cylinder confined to a brief period while the piston is in the region of bottom dead centre. Some external means of forcing fresh air into the cylinder is then necessary.

9.9.1 The Indicator Diagram

The diagram to the right of Fig. 9.9.1 shows schematically the variation in cylinder pressure with piston position in the working space of the engine; this diagram is reproduced to a larger scale in Fig. 9.9.3. The lines on this diagram are labelled to correspond with the four strokes of the cycle described in Fig. 9.9.2.

185

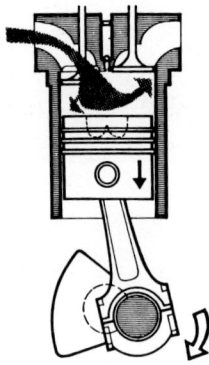

Induction stroke

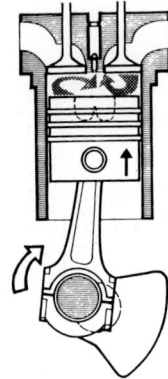

Compression stroke

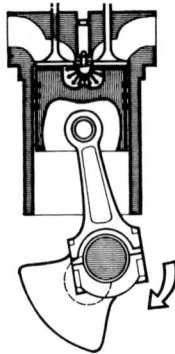

Combustion and Expansion stroke

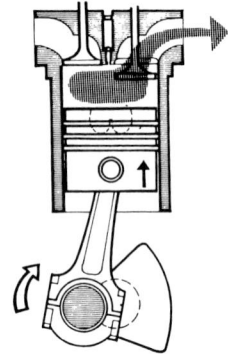

Exhaust stroke

Fig. 9.9.2 The four-stroke cycle

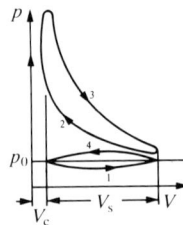

Fig. 9.9.3 Internal combustion engine indicator diagram

Stroke 1

Suction: Average pressure slightly less than atmospheric, corresponding to the pressure difference required to draw the fresh charge into the cylinder.

Stroke 2

Compression: Near the end of the inward movement of the piston ignition takes place and pressure rises rapidly.

186

Stroke 3

Expansion: Pressure continues to rise during the early part of the stroke as combustion continues. The cylinder charge then expands until just before bottom dead centre, when the exhaust valve opens and the pressure falls abruptly.

Stroke 4

Exhaust: During the inward movement of the piston the contents of the working space are discharged, requiring a small positive pressure in the cylinder.

An important measure of engine performance is the indicated power:

$$P_i = n \int p \, dV$$

where n = number of power strokes per unit time (one-half the number of engine revolutions for a 4-stroke engine) and the integral represents the area enclosed between the compression and expansion stroke of the indicator diagram. The indicated mean effective pressure (i.m.e.p.) is the height p_i of a rectangle of the same area as the indicator diagram. The indicated power output of the engine is thus the product of i.m.e.p., the swept volume V_s and the number of working strokes per unit time n:

$$P_i = p_i \cdot l \cdot A \cdot n = p_i V_s n \qquad (9.15)$$

where A = cross-sectional area of cylinder bore
    l = piston stroke
    V_s = swept volume of engine.

The concept of mean effective pressure is also extended to express the mechanical power output of an engine:

$$P_s = p_b . l . A . n \qquad (9.15a)$$

where P_s = power output
    p_b = brake mean effective pressure (b.m.e.p.).

Mechanical efficiency:

$$\eta_m = \frac{p_b}{p_i}$$

Typical values of b.m.e.p. at maximum power output would be 10 atm. for an automotive petrol engine and 8 atm. for a vehicle diesel engine. Supercharging permits these values to be doubled or trebled.

The much smaller area enclosed by the suction and exhaust lines may also be expressed as a mean effective pressure, and this corresponds to the "pumping losses" involved in charging and evacuating the cylinder.

Many instruments have been devised for recording the pressure changes in the cylinder of a running engine. The earliest, invented by James Watt, is still quite widely used for routine observation of the performance of low-speed marine diesel engines, and a current design of Watt indicator is shown in Fig. 9.9.4. A piston is subjected to the pressure in the engine cylinder by way of a tapping in the cylinder head or wall and acts against a calibrated spring. The movement of the piston is magnified by means of a linkage and moves a pencil which rests lightly against the surface of a paper

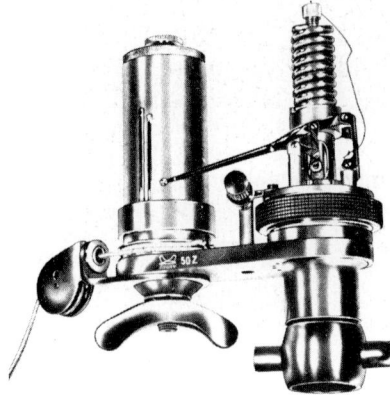

Fig. 9.9.4 Mechanical engine indicator

wrapped round a drum. The pencil moves parallel to the drum axis while the latter is rotated by a linkage driven from the engine which reproduces to an appropriate scale the piston movement. It will be apparent that the pencil will trace out a diagram which, when the appropriate scale factors are inserted, represents the *p-V* diagram for the process in the engine cylinder.

A great many ingenious designs of indicator have been proposed through the years, but the field is now effectively monopolized by the oscilloscope in which a signal from a transducer, typically a pressure-sensitive quartz crystal, is fed to the *Y*-axis of a cathode ray tube while the sweep of the *X*-axis is controlled by the engine crankshaft. The indicator thus displays a pressure–crank angle diagram rather than a *p-V* diagram—this in fact reveals far more information about the processes taking place in the cylinder, and in particular about the progress of combustion, than does the *p-V* diagram. Indicators are frequently fitted with electronic circuitry which permits conversion of the pressure–crank angle (*p-θ*) diagram to a *p-V* diagram, and Fig. 9.9.5 shows corresponding indicator diagrams for a petrol engine at full throttle.

It is possible, using a storage oscilloscope, to determine the indicated power of the engine by computation of the area of the indicator diagram, and reasonably accurate measurements of indicated power may be made in this way, particularly with large slow-running engines.

9.9.2 Air Standard Cycles

In this book we have repeatedly emphasized the value of ideal theoretical standards of efficiency with which the performance of real machines may be compared. We shall see later that classical thermodynamic theory has contributed little to the evolution of the internal combustion engine, but the various air standard cycles that have been proposed give at least a first approximation of the performance to be expected from a real engine.

We shall consider only one of these cycles: the constant volume or Otto cycle, Fig. 9.9.6. This cycle is based on the following assumptions:

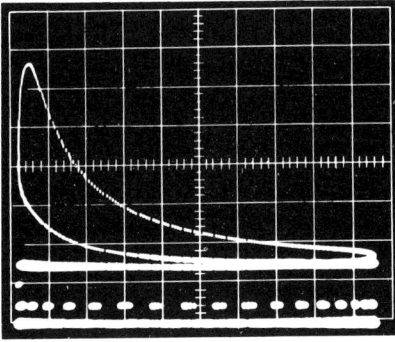

(a)

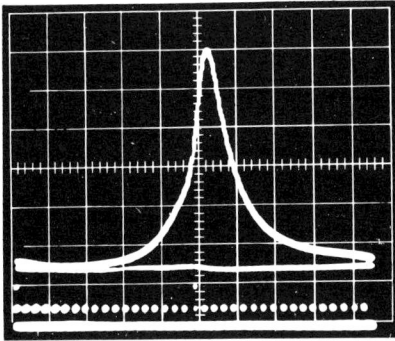

(b)

Fig. 9.9.5 $p\text{-}V$ (a) and $p\text{-}\theta$ (b) indicator diagrams for a petrol engine at full power

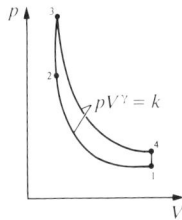

Fig. 9.9.6   The Otto or constant volume cycle

(a) The working fluid throughout the cycle is air, and this is treated as a perfect gas with invariant values of c_v and γ.

(b) The compression process 1–2 and the expansion process 3–4 are both adiabatic.

(c) In place of heat addition by internal combustion the phase 2–3 of the process is represented by the addition of heat from an external source, the volume of working fluid remaining constant during this process and the temperature T_3 corresponding to the maximum reached in the actual cycle.

189

(d) The exhaust process is replaced by cooling at constant volume from temperature T_4 to the original temperature T_1.

It will be evident that these conditions differ widely from those encountered in an engine; nevertheless the thermodynamic analysis of the cycle, which is very simple, gives certain useful indications regarding the performance of actual internal combustion engines.

$$\text{Efficiency} = \frac{\text{Power output}}{\text{Heat supplied}}$$

$$\eta_{as} = \frac{Q_{23} - Q_{41}}{Q_{23}}$$

$$= 1 - \frac{Q_{41}}{Q_{23}}$$

$$= 1 - \frac{mc_v(T_4 - T_1)}{mc_v(T_3 - T_2)}$$

$$\eta_{as} = 1 - \frac{T_4 - T_1}{T_3 - T_2} \tag{9.16}$$

For isentropic expansion and compression, Table 9.2.1:

$$\frac{T_2}{T_1} = \frac{T_3}{T_4} = r^{\gamma - 1}$$

$$\frac{T_2}{T_3} = \frac{T_1}{T_4}$$

$$\therefore 1 - \frac{T_2}{T_3} = 1 - \frac{T_1}{T_4}$$

$$\frac{T_3 - T_2}{T_3} = \frac{T_4 - T_1}{T_4}$$

$$\frac{T_4 - T_1}{T_3 - T_2} = \frac{T_4}{T_3}$$

whence:

$$\eta_{as} = \frac{T_3 - T_4}{T_3} \tag{9.17}$$

$$= 1 - \frac{1}{r^{\gamma - 1}} \tag{9.18}$$

The corresponding Carnot cycle efficiency, based on the maximum and minimum temperatures of the cycle, is:

$$\eta_{\text{Carnot}} = \frac{T_3 - T_1}{T_1}$$

Equation (9.17) indicates that the air standard cycle efficiency is less than the Carnot cycle efficiency since $T_4 > T_1$. Equation (9.18) makes it clear that the efficiency depends on the compression ratio only and not on either the initial temperature or on the amount of heat added in phase 2–3 of the cycle, Fig. 9.9.7(a), which may be regarded as the "indicator diagram" of the Otto cycle engine. The cycle would yield the same thermal efficiency whatever the i.m.e.p. The work ratio, however, increases with increased heat addition and greater i.m.e.p., and since in practice, as we have seen, higher efficiency is associated with higher work ratios, the highest practicable values of i.m.e.p. are desirable.

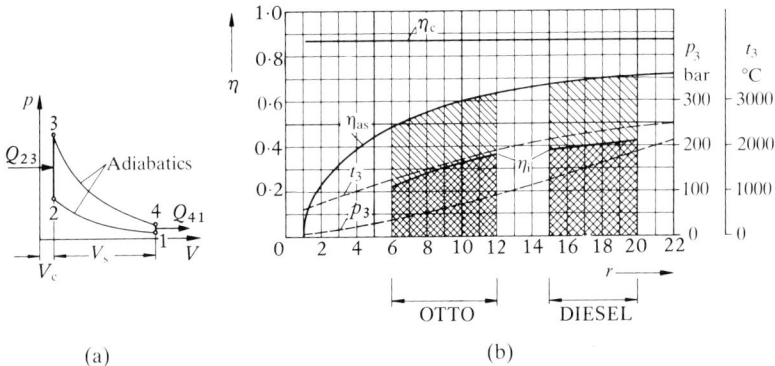

(a)   (b)

Fig. 9.9.7   Variation of air standard cycle efficiency with compression ratio

Fig. 9.9.7(b) shows the variation of air standard cycle efficiency with compression ratio, together with typical values of pressure p_3 and temperature T_3 at the end of combustion for initial conditions $T_1 = 300$K, $p_1 = 10^5$ N/m². The figure also indicates the range of compression ratios employed in petrol and diesel engines and the approximate indicated thermal efficiency η_i attainable in practice. It is clearly desirable to employ as high a compression ratio as possible in the petrol engine, in order to achieve maximum efficiency. The maximum usable compression ratio is determined primarily by the properties of the fuel and the onset of pre-ignition, see Experiment 12 (p. 202). The minimum compression ratio in a diesel engine is determined by the minimum value of the temperature T_2 and the corresponding charge density that is necessary to ensure reliable self-ignition of the fuel. The maximum compression ratio is limited by the maximum cylinder pressure that can be tolerated in the engine.

Many more complicated theoretical cycles have been proposed. These take into account such factors as heat addition, partly at constant volume and partly at constant pressure, and the variation of the specific heat of air with temperature. Complex computer programs taking into account the properties of the fuel, the exact course of the combustion process and effects such as dissociation are also available. However, while such analyses may provide useful guidance at the design stage, they are no substitute for the painstaking analysis of actual engine performance that is such a characteristic feature of the evolution of the internal combustion engine.

9.9.3 Engine Testing

The internal combustion engine is perhaps the best mechanical device available for introducing a student to the practical aspects of engineering. Our last two experiments, concerned respectively with a diesel engine and a petrol or spark ignition engine, are in a different class from those encountered so far in this book; they introduce the student to the kind of laboratory work that forms the staple occupation of industrial development departments. An engine is a comparatively complicated machine, occasionally alarming in its behaviour and capable of presenting many puzzling problems and mystifying faults, the solution of which can present a challenge even to the experienced engineer.

The safe management of an engine test installation calls for considerable skill and experience; careful instruction and guidance from the laboratory technician are essential. The operating handbook of the particular engine that is to be tested should be studied with care, and in particular the sections on preparing the engine for operation, running in, starting and stopping the engine, and lubrication and maintenance. All aspects of the installation: the design of the engine mountings and the method of coupling the engine to the dynamometer, the cooling water system, the exhaust system and the fuel handling system require careful attention. Reference [10] gives useful guidance in these matters.

The power output of an internal combustion engine is generally determined by coupling it to a dynamometer, a device which absorbs the engine power and permits measurement of the torque developed by the engine:

$$T = \frac{FL}{1000} \tag{9.19}$$

where $\quad T$ = engine torque, Nm
L = length of dynamometer torque arm, mm
F = force required to restrain torque arm, N.

The corresponding power output of the engine:

$$P_s = \frac{2N\pi \cdot T}{60} \quad \text{W}$$

where $\quad N$ = engine rev/min.

Fuel consumption is generally determined by observing the time t taken to consume a calibrated mass or volume of fuel, also recording the number of revolutions completed by the engine during this period, and the mean value of output torque T.

For a calibrated volume V_g l, fuel consumption V_f l/h is given by

$$V_f = \frac{3600 \, V_g}{t} \tag{9.20}$$

The specific fuel consumption of the engine is defined as

$$v_s = \frac{V_f}{1000 \, P_s} \quad \text{litres/kW hour} \tag{9.21}$$

or

192

$$m_s = \frac{f V_f}{1000\, P_s} \quad \text{kg/kW hour} \tag{9.22}$$

where f = density of fuel, kg/l.

The power output of an internal combustion engine is limited by its "breathing capacity"; there is never difficulty in introducing any desired quantity of fuel into the cylinder, but the maximum amount of fuel that can be burned depends entirely on the amount of air available for combustion. It follows that the volumetric efficiency of an engine is an important aspect of performance.

Volumetric efficiency:

$$\eta_{vol} = \frac{V_a}{V_s n} \tag{9.23}$$

where V_s = swept volume of each cylinder

$\quad\quad\quad\; n$ = number of working strokes per unit time

$\quad\quad\quad\; V_a$ = volume of air aspirated, at ambient conditions, per unit time.

Air consumption is commonly measured by drawing air through a sharp-edged orifice into an airbox coupled to the engine inlet and measuring the pressure drop across the orifice by means of a sensitive manometer. A certain minimum airbox capacity is necessary in order to damp out pressure pulsations. Recommendations regarding the use of orifices for air-flow measurement are given in [23] and criteria for determining airbox capacity in [24].

Exhaust temperature may be measured by a thermocouple in the exhaust pipe, with appropriate precautions regarding location and shielding (see Chapter 2), while the rate of heat transfer to the engine cooling system is determined by measuring coolant flow rate and temperature rise.

Many other measurements are made in the course of engine development, among which may be listed:

The analysis of exhaust gas composition.
Detailed studies of the combustion process.
Measurements of stress in the various components.
Studies of temperature distribution and heat flow.
Measurement of lubricant consumption.
Noise analysis.
Analysis of performance under transient conditions.

However, the basic measurements listed above, together with the indicator diagram, yield a vast amount of information regarding the performance and potential for improvement of any internal combustion engine.

The typical industrial engine test installation incorporates much more elaborate instrumentation than the simple methods to be described; computerized control systems, automatic programming and data logging are now commonplace. However, elaborate methods do not necessarily improve the accuracy of the results obtained; on the contrary elaborate instrumentation, if not carefully maintained and calibrated, may give rise to all kinds of unsuspected errors.

Fig. 9.9.8 Engine test bed

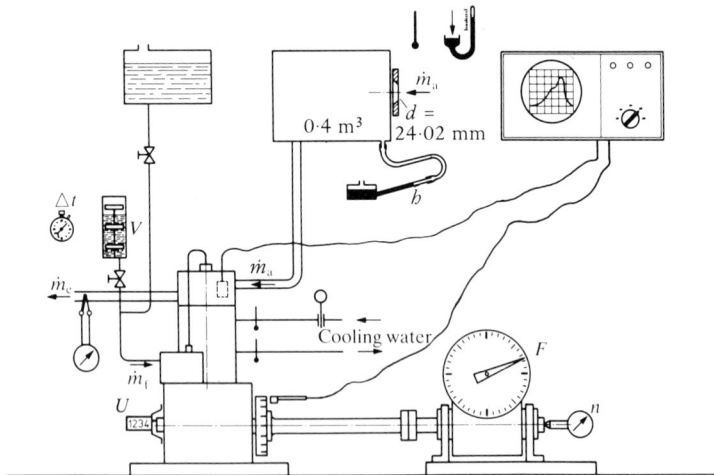

Fig. 9.9.9 Schematic arrangment of engine test bed

9.9.4 Experiment 11: Performance of a Small Diesel Engine

The test bed used by the authors is shown in Fig. 9.9.8, a schematic layout of the installation in Fig. 9.9.9 and a transverse section of the engine in Fig. 9.9.10.

The test procedure was first to warm up the engine and then to take a series of measurements of power output and fuel consumption covering the full range of engine performance. Speed was under the control of the engine governor and varied appreciably with load.

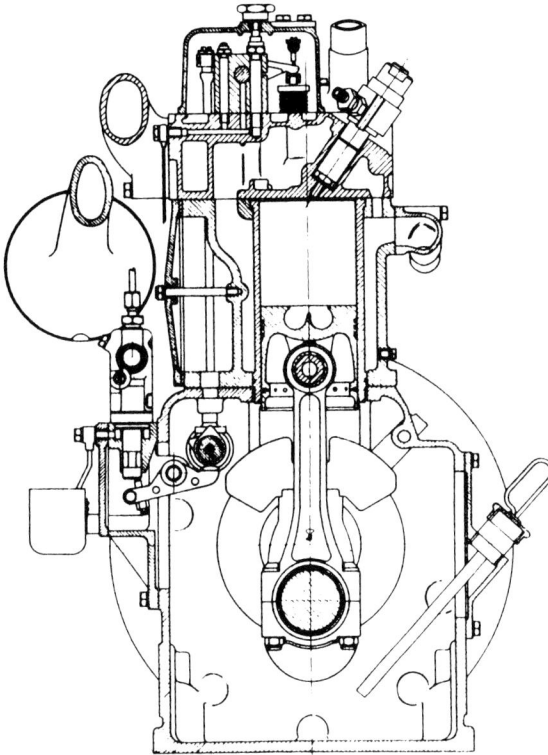

Fig. 9.9.10 Transverse section of four-stroke water-cooled diesel engine

A final set of observations was made in order to provide an estimate of the mechanical losses in the engine. The fuel supply is interrupted by means of the "stop" lever, and the dynamometer controls are manipulated so that the dynamometer functions as a motor, driving the engine at full speed. The corresponding torque is observed and the measurement is carried out as quickly as possible to minimize the influence of changes in engine temperature and lubrication conditions; these set in the moment the fuel supply is interrupted and result in changes in the level of mechanical loss.

9.9.5 Measurements and Calculations

Technical specifications of the authors' engine were as follows:

Single-cylinder, water-cooled, direct injection 4-stroke diesel engine.

Cylinder bore	87·3 mm
Piston stroke	110 mm
Swept volume	0·658 l
Compression ratio	15·5:1
Rated output	5·6 kW at 1800 rev/min.

Table 9.9.1 shows a set of test results, and differs in form from those given previously

195

Table 9.9.1 Experiment 11: Test Results

TEST SHEET — I.C. ENGINES

DATE	CUSTOMER		WORKS ORDER NO.	UNIT NO.
DYNAMOMETER TORQUE ARM	L = 220 mm	VOLTS	AMPERES	ARM. REG. RESISTANCE
ENGINE PETTER PHW 1	BORE 87.3 mm · STROKE 110 mm	CYLINDERS 1 · SWEPT. VOL. 659 cc	FUEL GASOIL · OIL FLEETOL 20/50	VARI. JET
BAROMETER 767 mmHg	AIR TEMP. 27 °C	AIR BOX SIZE	FUEL Litre/Hour = 4.195 · ORIFICE DIA. mm	FLOWMETER NO. · FUEL GAUGE

POWER: $W = \dfrac{F_N \, n}{43{,}40}$ b.m.e.p. = F_N kN/m² $\dfrac{90}{t}$ FUEL Litre/Hour = 4.195

	TACHO rev/min	COUNTER n rev	rev/min N	BRAKE LOAD F_N	POWER W	b.m.e.p. kN/m²	t sec 25/cc	FUEL litre/hour	FUEL litre/kW-hr	FUEL litre/hour (1800 rev/min)	
1	1600	806	1667	176	6750	738	29.0	3.10	0.459	3.35	
2	1700	1130	1703	165	6470	693	39.8	2.26	0.349	2.38	
3	1750	1224	1728	155	6170	651	42.5	2.12	0.343	2.20	
4	1750	1326	1729	145	5770	609	46.0	1.95	0.339	2.03	
5	1790	1558	1758	123	4980	516	53.2	1.69	0.339	1.73	
6	1800	1852	1788	101	4160	424	62.2	1.44	0.346	1.44	
7	1800	2148	1822	79	3310	331	70.8	1.27	0.383	1.25	
8	1800	2432	1820	66	2760	277	80.2	1.12	0.406	1.10	
9	1805	2604	1800	55	2280	231	86.8	1.03	0.451	1.03	
10	1810	3240	1838	34	1430	142	105.8	0.85	0.594	0.83	
11	1820	4254	1866	12	510	50	136.8	0.65	1.29	0.62	
12	1820	3722	1846	25	1060	105	121.0	0.74	0.70	0.72	
13	1720	1732	1732	−64	−2550	−268	60.0	—	—	—	Motoring test

in that it shows a test sheet typical of those used in an industrial testing or development department for recording experimental data and the subsequent calculations.

Test point No. 2 is analysed below:

Brake arm	$L = 220$ mm
Fuel gauge volume	$V_g = 0.025$ l
Fuel consumption rate	$V_f = \dfrac{90}{t}$

where $t =$ time to consume calibrated volume

Density of fuel	$\rho_f = 0.850$ kg/l

The test sheet shows the number of revolutions n made by the engine in time t. Then:

Engine speed	$N = \dfrac{60_n}{t}$ rev/min
Engine power output	$P_s = \dfrac{NF}{43 \cdot 40}$ W

The brake mean effective pressure, see equation (9.15a), is derived from the power output:

$$p_b = \frac{120 \, P_s}{V_s N}$$

noting that n in equation (9.15a) equals $\frac{1}{2} N/60$ since in a four-stroke engine there is only one working stroke per two revolutions.

For test point 2:

$$F = 165 \text{ N}$$
$$t = 39 \cdot 8 \text{ s}$$
$$n = 1130 \text{ rev.}$$
$$N = 1703 \text{ rev/min}$$
$$P_s = 6470 \text{ W}$$
$$p_b = 693 \text{ kN/m}^2$$
$$V_f = 2 \cdot 26 \text{ l/h}$$
$$v_s = 0 \cdot 3491 / \text{kW-h}$$
$$m_s = 0 \cdot 297 \text{ kg/kW-h}$$
$$\text{corrected } V_f = 2 \cdot 39 \text{ l/h}$$

The last figure is the fuel consumption rate adjusted proportionately to a speed of 1800 rev/min; the use of this figure is explained below.

The thermal efficiency is calculated taking a Lower Calorific Value:

$$\text{Net} = 43\,000 \text{ kJ/kg}$$
$$\eta_{th} = \frac{3\,600\,000}{43 \times 10 \times 0 \cdot 297} = 0 \cdot 282$$

Fig. 9.9.11 shows some of the more important characteristics of the engine. The "Willans line" shows the relation between corrected fuel consumption and b.m.e.p. Since the engine speed is constant, this is also, to a different scale, a curve of fuel

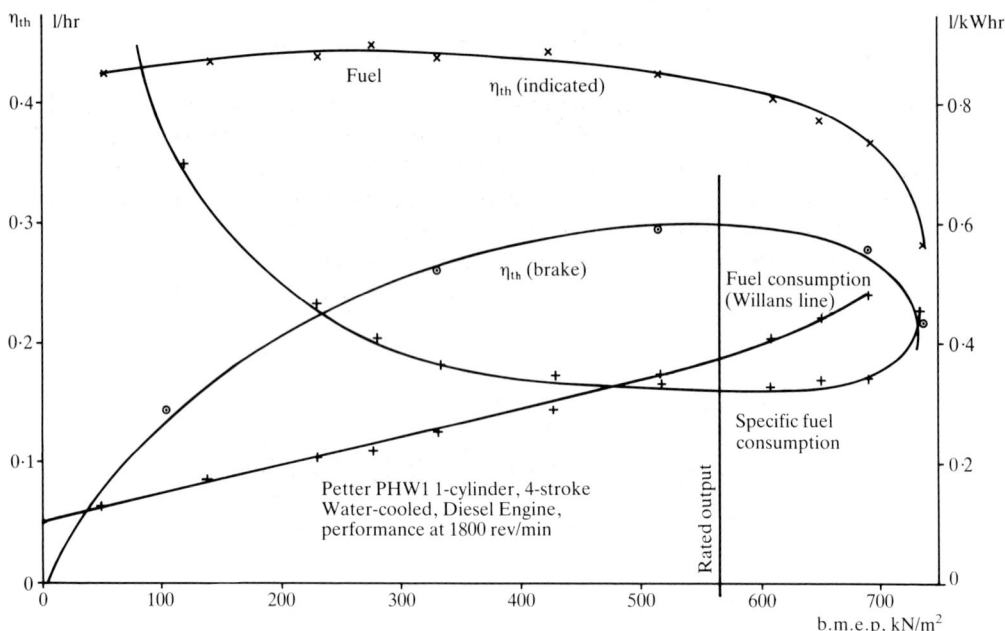

Fig. 9.9.11 Performance curves for diesel engine. Relation between fuel consumption, specific fuel consumption and b.m.e.p.

consumption against power output. The curve is of characteristic form for a diesel engine, with the test points lying almost exactly on a straight line up to about 80 percent of rated power output. This provides the basis for another method of estimating the mechanical losses in the engine: it is a reasonable assumption that in the straight-line part of the diagram, indicated power output is directly proportional to fuel consumption; hence, by extrapolating the straight line to zero fuel consumption we may make an estimate of the friction mean effective pressure (f.m.e.p.) corresponding to the power required to overcome the mechanical and pumping losses in the engine. Fig. 9.9.11 indicates an f.m.e.p. of 222 kN/m².

Test point 13 gives a value for friction mean effective pressure, taken by motoring the engine immediately after shut-down, of 268 kN/m². It is usual to find that a measurement of mechanical losses taken by this method yields a value slightly higher than the true one, mainly because the cylinder wall temperatures are lower than when the engine is firing, resulting in greater viscous friction, and because the absence of combustion results in an increase in the pumping losses, as a consequence of the change in flow pattern in the exhaust system. In calculating the mechanical efficiency, a mean value for f.m.e.p. = 245 kN/m² has been assumed.

Taking this value, we may estimate the i.m.e.p. for the test point analysed above:

$$p_b = 693 \text{ kN/m}^2$$
$$p_i = 693 + 245 = 938 \text{ kN/m}^2$$

whence

$$\eta_{mech} = \frac{693}{938} = 0.74$$

and indicated thermal efficiency

198

$$\eta_{th(i)} = 0.282 \cdot \frac{938}{693} = 0.38$$

Fig. 9.9.12 shows an indicator diagram taken at full power output on a crank angle basis. Ignition delay in a diesel engine is the interval between the start of injection of the fuel into the cylinder and the first detectable pressure rise on the indicator diagram. It is a measure of the time required for the initiation of combustion by the transfer of heat from the hot compressed air in the cylinder to the fuel droplets and the subsequent rate of pressure rise, which determines the degree of roughness or smoothness in the running of the engine, depending critically on the length of the ignition delay.

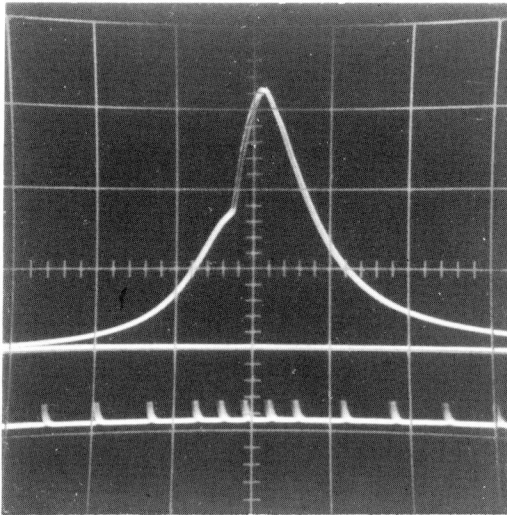

Fig. 9.9.12 p-θ indicator diagram for diesel engine at full power

The start of injection may be determined by means of a "needle lift diagram". A transducer coupled to the oscilloscope senses movement of the needle valve of the fuel injector, enabling the instant corresponding to opening of the injector to be determined. The ignition delay is typically about 10° of crank angle or, at 1800 rev/min, an interval of rather less than 1 millisecond—an indication of the rapid rate at which processes take place in the internal combustion engine.

9.9.6   Discussion of Results

The compression ratio of this engine is 16·5:1. Inserting this value in equation (9.18) gives a theoretical indicated thermal efficiency of $\eta_i = 0.674$ compared with an actual efficiency of 0·38.

It has already been pointed out that the theoretical equation for air cycle efficiency takes no account of various factors, among them the chemical dissociation of the combustion gases and the increasing specific heat of air at elevated temperatures. A modification to equation (9.18) has been proposed [9] to take into account these

199

effects. This involves a reduction in the index of compression and expansion from its theoretical value of $\gamma = 1 \cdot 4$ to an effective value of $1 \cdot 34$. This yields a modified version of equation (9.18):

$$\eta_{as \text{ (modified)}} = 1 - \frac{1}{r^{0 \cdot 34}} \tag{9.24}$$

and a corresponding modification to the theoretically obtainable efficiency of the present engine:

$$\eta_{i \text{ (modified)}} = 0 \cdot 614$$

The actual indicated thermal efficiency is thus about 72 percent of the theoretical value. The ideal thermal efficiency is calculated on the assumption that all the heat of combustion not converted into mechanical work, namely 38·6 percent of the total, is rejected in the exhaust gas. It may be shown, however, that in the actual engine the heat rejected in the exhaust gas amounts to only about 25 percent of the heat of combustion, while some 45 percent of the heat of combustion is rejected either to the cooling water or to the surroundings. This gives a clue to one of the main reasons for the shortfall in the performance of the actual engine: during the compression, combustion and expansion phases of the cycle intensive heat loss takes place from the working fluid to the walls of the cylinder and combustion chamber and to the piston surfaces, resulting in a depression in the temperature and hence in the pressure achieved in the cylinder at all points in the cycle when compared with the corresponding theoretical values. Much effort is currently being devoted to the development of the so-called "adiabatic engine" in which the surfaces of cylinder liner, cylinder head, valves and piston crown are of ceramic, permitting very much higher temperatures and a reduction in this loss.

A further reason for the defect in engine performance may be deduced from Fig. 9.9.12. This shows that the cylinder pressure, instead of rising abruptly from that corresponding to the end of the compression stroke to its maximum value at top dead centre, begins to rise sometime before t.d.c. and is still rising, to reach a maximum about 5° after t.d.c. This results in both an increase in the compression work and a reduction in the expansion work available. It may be shown further that combustion, with the assumptions of the ideal cycle complete at top dead centre, in fact persists until quite late in the expansion stroke. The effect of these further departures from the conditions assumed in calculating the air cycle efficiency is a further substantial reduction in the area of the indicator diagram.

The mechanical efficiency of the engine, approximately 74 percent, is a typical value for a small diesel engine and shows that rather more than one-quarter of the power developed in the cylinder is lost in the form of mechanical friction, power absorbed by ancillary equipment, and in pumping losses. It should be noted that a part of these mechanical losses appear as heat in the cooling water.

Fig. 9.9.11 (p. 198) illustrates one of the good features of the diesel engine: the thermal efficiency remains fairly constant between about one-third and full power output. It will be seen later that the petrol engine has a less favourable performance in this respect.

The authors' engine shows a maximum thermal efficiency, based on power output, of 28·2 percent. Very large, slow-running marine diesel engines, which may have a bore of 900 mm, a stroke of 2600 mm and a swept volume approaching two thousand times that of the authors' engine, have achieved thermal efficiencies of over 50 percent. This is an example of scale effect, which invariably favours large machines.

9.10 The Performance of the Petrol Engine

An understanding of the combustion process in a spark ignition engine involves a synthesis of practically all the material that has formed the subject-matter of this book so far, as is clear from a consideration of the processes involved:

(a) The formation of a mixture of gas and vapour by the evaporation of liquid droplets of fuel in an airstream (Chapters 6 and 7).
(b) The initiation of combustion by an electric spark (Chapter 8).
(c) The propagation of a flame through a volume of combustible mixture (Chapter 8 and reference to the propagation of pressure- and shock-waves).
(d) The release of heat with corresponding increase in temperature and pressure, and a change in the bulk properties of the working fluid as this is transformed from a mixture of air and fuel to a volume of combustion products (Chapters 2, 5, 6 and 8).
(e) Heat transfer by radiation, natural and forced convection from the working fluid to its surroundings (Chapter 3).
(f) The performance of mechanical work by the expansion of the products of combustion (Chapter 6).
(g) The energy balance of the complete process (Chapter 4).
(h) Comparison with the ideal attainable performance (Chapter 5).

It is clear that we are dealing here with a large number of interlocking phenomena, involving the entire field of mechanics of fluids, heat transfer, classical and chemical thermodynamics. It must, however, be pointed out again that the contribution made by theory to the understanding of the internal combustion engine and to its development has been minimal.

It would go far beyond the limits of the book to discuss in any detail the combustion process in a spark ignition engine, but perhaps a brief account would not be out of place. The mixture of fuel and air enters the cylinder from the carburettor as what has been described [9] as a "shower of rain". By the end of the compression stroke the fuel will have been evaporated to form a reasonably homogeneous charge of combustible gas, at a temperature of perhaps 250–350°C and pressure of perhaps 25 atmospheres.

A spark is passed between the points of the sparking plug and if the process is observed through a quartz window in the combustion chamber it is seen that for an appreciable period after the passage of the spark a small sphere of flame persists in the neighbourhood of the points. After an interval, which is not exactly constant in terms of crank angle degrees from one stroke to the next, this sphere of flame begins to grow and to travel across the combustion chamber. With suitable optical methods it is possible to observe that the flame front does not travel uniformly but is broken up by the turbulence in the chamber. Eventually, if combustion proceeds normally, the flame will have traversed the whole of the combustion chamber and will be extinguished before the piston has travelled an appreciable distance on its expansion stroke.

A great part of the research and development effort devoted to the internal combustion engine has had for its object the understanding of this process and of the very numerous factors that affect it. These include:

(a) the ratio of air to fuel;
(b) the chemical properties of the fuel;
(c) the compression ratio of the engine;
(d) the form of the combustion chamber;
(e) the nature of the air motions (turbulence) in the combustion chamber;
(f) the degree of cooling of the chamber walls, cylinder bore and piston crown;
(g) the location of the sparking plug and of the valves, in particular of the hot exhaust valve;
(h) the timing of the ignition spark.

A petrol engine is capable of operating on a range of air-to-fuel ratios (mixture strengths) by weight from about 8:1 to 22:1. The former corresponds to a "rich" mixture (excess fuel) and the latter to a "weak" mixture (excess air). The general performance of the engine in terms of power output, economy, production of pollutants in the exhaust, and general smoothness of operation depends critically on the mixture strength.

It is the function of the carburettor to regulate the mixture strength to the most suitable value over the full engine operating conditions. These involve, in an automobile engine, the full speed-range, the full range of load, acceleration overrun and idling conditions. These requirements make very exacting demands upon the carburettor and induction system, and explain the reason for the complicated nature of modern carburettor design.

An alternative solution is to inject the fuel, either into the manifold or into the individual inlet ports, using an electronically controlled fuel injection pump. This system is in principle capable of better results than even the most elaborate carburettor, but at the cost of considerable complication.

9.10.1   Experiment 12: Performance of a Variable Compression Ratio Petrol Engine

An overall understanding of the characteristics of the petrol engine is much more easily achieved if the engine is modified to permit variation of two of the principal parameters: the compression ratio and the mixture strength. A number of variable compression-ratio engines are available, some of them elaborate and expensive units intended for the rating of fuels and for research purposes.

Fig. 9.10.1 and 9.10.2 show a very simple engine of this kind employed by the authors. This is a single-cylinder 4-stroke unit having a cylinder head of special design incorporating a counter-piston, the position of which may be varied while the engine is running to give a range of compression ratios from 4:1 to 10:1. The typical modern automobile engine has a compression ratio towards and a little beyond the top end of the range of the authors' engine.

Mixture strength may be varied relatively simply by the use of a carburettor fitted

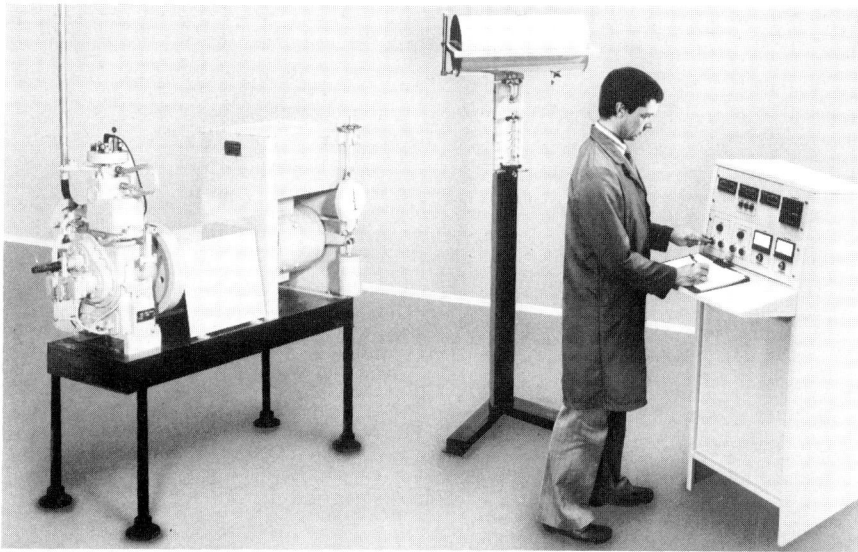

Fig. 9.10.1 Variable compression-ratio petrol engine test bed

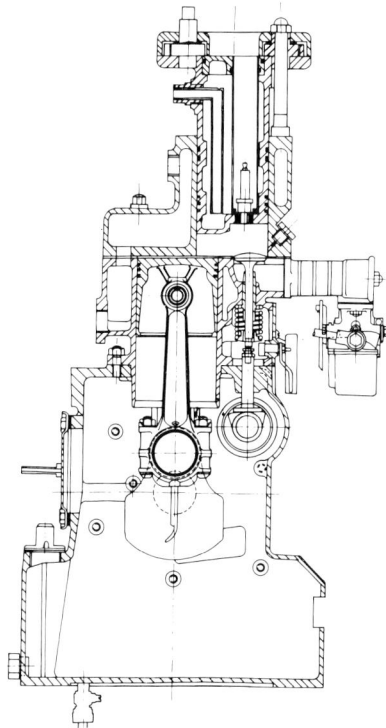

Fig. 9.10.2 Transverse section of variable compression-ratio petrol engine

with a variable jet. This consists merely of a needle valve associated with the main fuel-supply jet of the carburettor, by adjustment of which the amount of fuel admitted to the airstream may be varied and the mixture strength altered at will. The engine is coupled to an electrical dynamometer and the range of instrumentation is essentially similar to that employed for the diesel engine test of Experiment 11, with the addition of a sharp-edged orifice and airbox for the measurement of air consumption.

9.10.2 Measurements and Calculations

The test procedure involves running the engine at full load and constant speed and at a series of different compression ratios. For each compression ratio the mixture strength is varied by means of the variable jet over the full range at which the engine can operate. The test is started by adjusting the mixture strength so that the engine develops maximum power. Subsequent measurements will show that this corresponds to an air/fuel ratio of about 12:1 by weight, corresponding to a rather rich mixture; the stoichiometric air/fuel ratio is about 14·5:1 for petrol.

The mixture is now made richer by opening the needle valve further, when it is found that the power of the engine decreases slightly while the fuel consumption, of course, increases. When the mixture becomes very rich the engine begins to run unsteadily and explosions may also take place in the exhaust system.

We now return to the mixture strength corresponding to maximum power and start to weaken the mixture, when it is found that the power of the engine begins to fall, at first slowly and then more rapidly. Eventually "popping back" takes place (explosions occur in the carburettor), the running of the engine again becomes uneven and eventually it stops.

Table 9.10.1 shows a set of observations taken at one compression ratio; it is appropriate, when a number of groups of students are available, to arrange for each group of take a set of readings at one compression ratio. The observations are best presented by a plot of the so-called "hook curve", Fig. 9.10.3, a curve of specific fuel consumption against mean effective pressure. Fig. 9.10.4 shows a series of hook curves taken at different compression rates. If, as in the present case, the air consumption is also measured, we may plot power output and specific fuel consumption against air/fuel ratio, Fig. 9.10.5.

9.10.3 Discussion of Results

Some experimental engines are fitted with a quartz window in the combustion chamber through which the combustion process may be observed. If such a window is fitted to an engine that is being taken through the above sequence of tests we see the following changes.

(a) At mixture strength corresponding to maximum power and over a range of weaker mixtures, combustion takes place smoothly and rapidly with a blue flame which is extinguished fairly early in the expansion stroke.
(b) As we proceed towards richer mixtures the combustion takes on a yellow colour, arising from incandescent carbon particles, and persists for a longer period, even being incomplete when the exhaust valve opens at the end of the expansion stroke.

Table 9.10.1 Experiment 12: Test Results

TEST SHEET — I.C. ENGINES

UNIT NO. TE 15/1864

DATE		CUSTOMER								WORKS ORDER NO.

DYNAMOMETER TORQUE ARM — ARM. REG. RESISTANCE

ENGINE	VARIABLE COMP.	BORE 85 mm	STROKE 82.5 mm	VOLTS 265 mm	AMPERES	CYLINDERS 1	SWEPT. VOL. 468 cc	FUEL SHELL 4-STAR	OIL SHELL 20W30	VARI. JET YES
BAROMETER 755 mmHg	AIR TEMP. 21 °C	AIR BOX SIZE				FUEL Litre/Hour = 150 litre	ORIFICE DIA. 18.14 mm	FLOWMETER NO.		FUEL GAUGE 'O'

POWER:kW = $\dfrac{F_N n}{36{,}000}$ b.m.e.p. = $7.123\,F_N$ kN/m²   FUEL Litre/Hour = $\dfrac{180}{t}$ NOTE: COMPRESSION RATIO 7 : 1

TACHO rev/min	COUNTER rev (2x)	COUNTER sec	rev/min n	BRAKE LOAD F_N	POWER kW	b.m.e.p. kN/m²	t sec 50/cc	FUEL litre/hour	FUEL litre/kW·hr	EXH. °C	HEAD cmH₂O h_o	TEMP °C T_A	VOL F/R l/s V_a	EFFY η VOL	MASS F/R g/s m_a	a/f RATIO	REMARKS
RICH 2000	788		2012	50.5	2.822	360	47.0	3.830	1.357	620	6.15	21	4.92	0.627	5.89	7.38	
	834		2002	67.0	3.726	477	50.0	3.600	0.966		6.15		4.92	0.630	5.89	7.85	
	891		1922	68.0	3.630	484	52.5	3.429	0.945	580	6.13		4.91	0.655	5.88	8.23	
	962		2025	71.0	3.993	506	57.0	3.158	0.791		6.10		4.90	0.620	5.87	8.92	
	1044		1989	74.0	4.089	527	63.0	2.857	0.699		6.05	21.5	4.88	0.629	5.84	9.81	
	1156		2010	74.5	4.160	530	69.0	2.609	0.627		6.00		4.86	0.620	5.82	10.71	
	1304		2019	74.5	4.178	530	77.5	2.323	0.556		5.97		4.85	0.616	5.80	11.99	
	1355		1995	74.5	4.129	530	81.5	2.209	0.535	675	5.93		4.83	0.621	5.79	12.58	
	1560		2035	69.0	3.900	491	92.0	1.957	0.502	685	5.93	19	4.83	0.609	5.79	14.20	
	1615		1988	65.2	3.601	464	97.5	1.846	0.513		5.87		4.81	0.620	5.76	14.98	
	1698		2008	60.0	3.347	427	101.5	1.773	0.530	673	5.95		4.84	0.618	5.80	15.70	
	1792		2010	52.0	2.903	370	107.0	1.682	0.580		6.03		4.88	0.623	5.84	16.66	
WEAK	1908		2000	35.0	1.944	249	114.5	1.572	0.809	682	6.03	19	4.88	0.626	5.84	17.83	

DENSITY OF FUEL 0.75 kg/litre

205

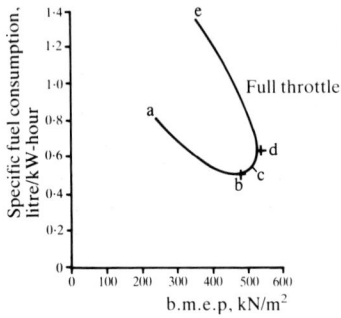

Fig. 9.10.3
The "hook" curve, relation between specific fuel consumption and b.m.e.p. for variable compression-ratio petrol engine

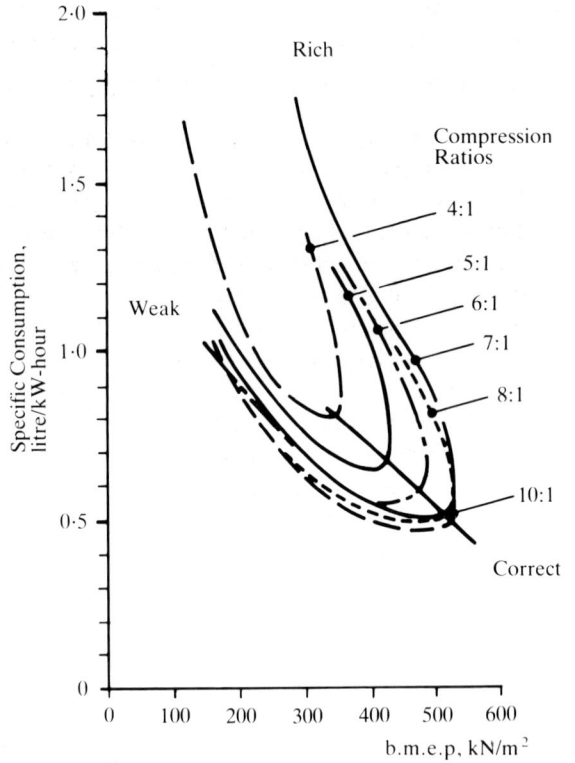

Fig. 9.10.4
Hook curves for variable compression-ratio petrol engine

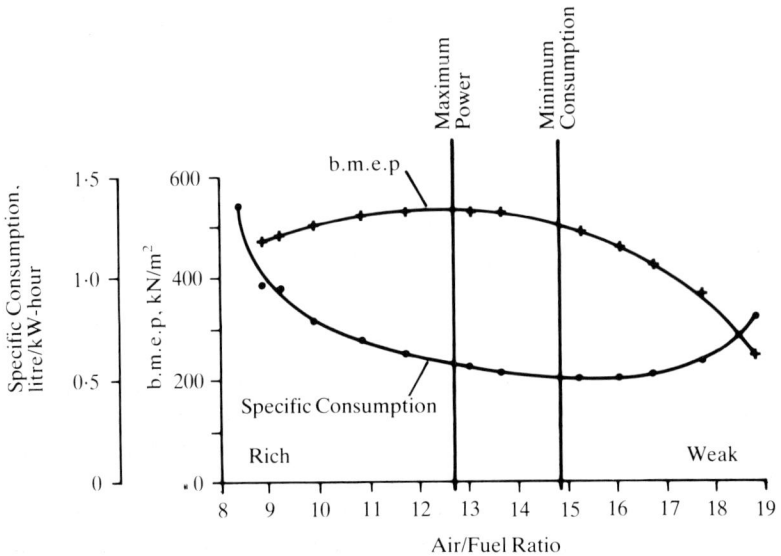

Fig. 9.10.5   Relation between b.m.e.p., specific fuel consumption and air/fuel ratio

206

(c) As we weaken the mixture the combustion retains a bluish colour, becomes more uneven and persists throughout the expansion stroke.

The following features of Fig. 9.10.3 call for comment:

(a) Point *a* corresponds to the weakest mixture at which the engine will run. The power output is about half the maximum, and the specific fuel consumption can be as much as twice that corresponding to the best efficiency.
(b) Point *b* corresponds to the best performance of the engine (maximum thermal efficiency). The power output at this point is about 95 percent of that at point *d* corresponding to maximum power.
(c) Point *c* corresponds to the stoichiometric air/fuel ratio.
(d) Point *d* corresponds to the maximum power output and the specific fuel consumption is then about 10 percent greater than at the point of best efficiency.
(e) Point *e* corresponds to the maximum mixture strength at which the engine will run. The power output is about 90 percent of maximum, and the specific fuel consumption is between 50 percent and 100 percent greater than that corresponding to the best efficiency point.

Without examining the matter further, it is evident that the correct functioning of the carburettor, which has for its duty the supply of the most suitable mixture of air and fuel to the engine, is of the utmost importance in achieving satisfactory performance both in terms of power output and fuel economy.

A comparison of the various hook curves shown in Fig. 9.10.4 shows that as we increase the compression ratio the performance of the engine improves; the minimum specific fuel consumption at a compression ratio of 10:1 is about half that at a compression ratio of 4:1, while the power output is almost doubled.

Fig. 9.10.5 shows a characteristic feature of petrol engine performance: the maximum power output corresponds to a slightly rich mixture, air/fuel ratio 12·7:1, while best economy corresponds to a slightly weak mixture, air/fuel ratio 14·9:1. These air/fuel ratios are disposed on opposite sides of the stoichiometric ratio.

9.10.4 Further Experiments and Questions

The internal combustion engine offers almost unlimited possibilities for experimental and development projects. It is noteworthy that after nearly a century of development, in the course of which more experimental effort has been devoted to the internal combustion engine than to any other mechanical device, laboratories all over the world are still occupied in attempting to improve the performance of this machine.

External circumstances are responsible for promoting much of this effort: some years ago governments, led by the United States, started to introduce legislation restricting the permitted level of exhaust-pollution discharged into the atmosphere by vehicle diesel and petrol engines. The attempt to meet these regulations, which has still by no means succeeded, has required a vast and continuing volume of development work.

In the early 70's the sudden increase in the cost of oil fuels prompted a much more intensive study than had been made in the past of the factors affecting engine fuel consumption, together with renewed efforts to develop the diesel engine in a suitable form for passenger cars. This has resulted in a fundamental re-examination of all aspects of engine design and performance, rendered more complex by the circumstance that the requirements of high thermal efficiency and low exhaust-pollution conflict to a considerable degree. Reference [10] gives some indication of the scope of experimental work applicable to an internal combustion engine.

10

Amplification of Some Earlier Sections

10.1 Quasi-static and Real Processes

The question arises: is it acceptable to represent the processes taking place in real thermodynamic machines on diagrams of state intended to represent equilibrium conditions? As an example we consider the indicator diagram of an internal combustion engine. This diagram represents the cyclic variation of pressure experienced by a transducer located at one point in the combustion chamber wall, typically at the end of a short passage. At high engine speeds, when the entire working cycle occupies only a few milliseconds, are we justified in assuming that the indicated pressure is a fair measure of the average pressure in the working space and of the pressure exerted by the gas on the piston?

We know from the subject of fluid mechanics that pressure changes in a fluid are propagated with the velocity of sound, something over 300 m/s in the conditions of an engine combustion chamber. Significant pressure differences within the working space (apart from high-frequency oscillation associated with resonant pressure waves) could thus only arise if the piston velocity were of comparable magnitude to the velocity of sound. Normally piston velocities do not exceed about 20 m/s, showing that we are in fact justified in neglecting pressure differences within the working space.

However, temperature differences are propagated in a gas very much more slowly than pressure differences; temperature differences of many hundreds of degrees may be present in the working space of an internal combustion engine. We are thus not justified in ascribing a particular temperature to the working fluid at the part of the cycle corresponding to the combustion and expansion processes; indeed we cannot even define the meaning of temperature in these rapidly changing conditions. This is particularly the case when combustion is taking place, since the composition of the gas and the distribution of energy between the different modes of motion of the gaseous molecules (see p. 28) are both indeterminate.

It follows that, strictly speaking, we are not justified in representing cyclic processes of this kind on a diagram of state. This is sometimes indicated by representing such non-equilibrium conditions by dotted lines. Real processes are, strictly speaking, always a succession of non-equilibrium states, since the condition of the system can only be changed by a disturbance of its equilibrium. A change of state that could be strictly represented by a series of equilibrium conditions is in principle reversible: an infinitesimal change of state can by definition take place in either direction.

Many technically important processes, and in particular steady-flow as opposed to cyclic processes, may, however, be represented with acceptable accuracy as a succession of equilibrium conditions. Such processes may legitimately be represented in diagrams of state and are known as quasi-static processes.

The expansion of a gas in a nozzle, Fig. 10.1.1(a) and its representation on the *T-s* diagram, Fig. 10.1.1(b) may be taken as an example. In the core of the gas stream the flow is very nearly perfectly frictionless and isentropic: it may legitimately be represented by 1–2 on the *T-s* diagram. In the boundary layer, however, the kinetic energy generated in the course of the expansion is almost entirely reconverted into heat as a consequence of friction, some of this heat being transferred to the nozzle wall. The process in the boundary layer may be represented by a line such as 1–3 on the *T-s* diagram.

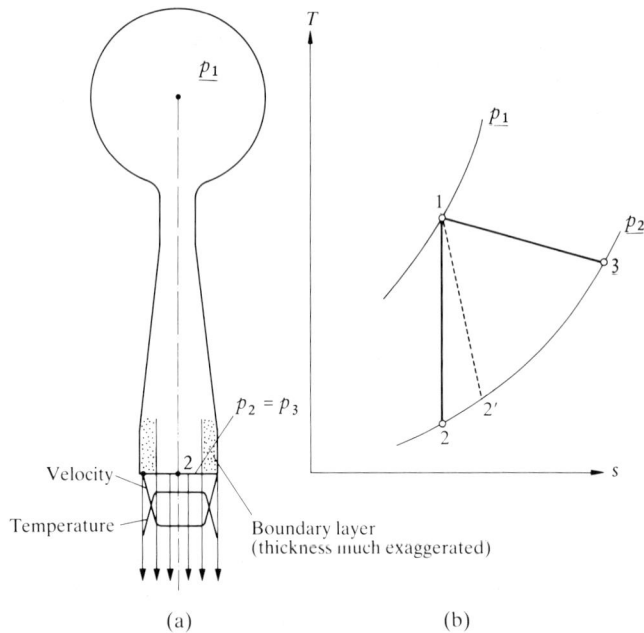

Fig. 10.1.1. (a) Adiabatic expansion of gas in a nozzle
(b) Representation on the *T-s* diagram

The total flow may be represented on the *T-s* diagram by a line such as 1–2′ representing the weighted average of the flow in the core and the flow in the boundary layer, but it will be apparent that the succession of states represented by this line has no specific physical meaning and it is appropriate to represent it by a dotted line.

10.2 The Steady-flow Energy Equation: Modification to take into account Potential and Kinetic Energy

So far in this book the steady-flow energy equation (4.5) has been expressed in a form that takes into account only internal energy, heat, and mechanical work. In

210

some cases a significant proportion of the energy of the working fluid is present in the following forms:

Kinetic energy: $\dfrac{mv^2}{2}$

Potential energy: mgz

where $v =$ velocity of working fluid

$z =$ position of working fluid in terms of its height above some reference level

Taking these forms of energy into account, the steady-flow energy equation is rewritten:

$$h_1 - h_2 = w_s - q + \frac{v_2^2 - v_1^2}{2} + g(z_2 - z_1) \tag{10.1}$$

The two additional terms in the equation represent, respectively, the difference between the kinetic energy of fluid leaving the system and that at entry and the difference between the datum levels at exit and entry. In the case of gases, the potential energy term may almost always be neglected; at the other extreme, in the application of the steady-flow energy equation to a water turbine, w_s and the potential energy term are the principal elements in the equation.

An important application of the extended steady-flow energy equation refers to the frictionless adiabatic expansion of a gas:

$$h_1 - h_2 = \frac{v_2^2 - v_1^2}{2} \tag{10.2}$$

From this equation we may derive a very simple expression for the velocity attained by a gas when expanding under these conditions from rest:

$$v_2 = v = \sqrt{2\Delta h} \tag{10.3}$$

On the assumption that the expansion is isentropic and frictionless as well as adiabatic, then if we know p_1, T_1 and p_2, the velocity at the end of the expansion may be derived very simply from the h-s diagram.

In the case of an ideal gas having constant specific heat, equation (10.2) may be written:

$$c_p(T_1 - T_2) = \frac{v_2^2 - v_1^2}{2} \tag{10.4}$$

An interesting application of equation (10.4) is in the calculation of stagnation temperatures. Fig. 10.2.1 represents the flow of air at velocity v_1 past a model in a wind tunnel. It is an experimentally observed fact that the pressure and temperature at the stagnation point agree with the value predicted from equation (10.4), on the assumption that the compression process accompanying the reduction of the air velocity to zero is an exact reverse of the expansion process of Fig. 10.1.1. This permits us to calculate the difference between the temperature at the stagnation point and in the main airstream:

$$T_2 - T_1 = \frac{v^2}{2c_p} \qquad (10.5)$$

The increase in temperature thus depends in a very simple way on the velocity. We have assumed only adiabatic conditions but not the absence of friction. The rapidity of the process and the comparatively moderate temperature gradients normal to the

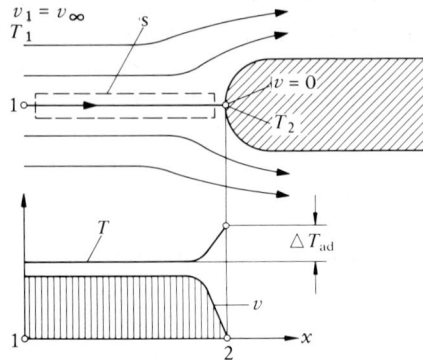

Fig. 10.2.1 Flow of air past a fixed object: development of stagnation temperature

direction of flow ensure that there is virtually no heat exchange with the surroundings. Equation (10.5) is also applicable to shock-waves associated with supersonic flow, giving a correct indication of the relative temperatures before and after the shock. For air, $c_p = 1$ kJ/kg K approximately, giving:

$$T_2 - T_1 = \frac{v^2}{2000} \qquad (10.6)$$

An air speed of 100 m/s corresponds to a temperature rise of about 5K, while at the speed of the Concorde aircraft, roughly 700 m/s, the temperature rise is some 220K. Note that the adiabatic temperature rise is independent of pressure and thus remains unchanged for an aircraft flying at great height. This is because c_p is virtually independent of pressure.

At positions on the surface of the body remote from the stagnation point the pressure rise will differ from that at the stagnation point, but roughly the same increase in temperature will occur. It will be appreciated from this discussion that at high fluid velocities it is quite impossible to determine the true temperature of a fluid stream by means of a stationary thermometer. Depending on the exact nature of the heat exchange between such a thermometer and its surroundings it will indicate a temperature somewhere between the stagnation temperature and the true temperature of the airstream.

At high fluid velocities we must thus distinguish between static and total temperatures and pressures. The static temperature and pressure are those observed in a fluid at rest or by an observer travelling with the fluid. The total temperature and pressure are those observed at the stagnation point on the surface of a body at rest in the fluid.

Experiments in fluid mechanics show us that we may readily measure the static pressure in a fluid stream by means of a pressure tapping in a wall the surface of which lies parallel to the direction of fluid flow: the pressure of the main fluid stream is transmitted to the wall through the boundary layer without alteration. This is not the case with the temperature, which rises very abruptly through the thickness of the boundary layer.

The analysis of the results of Experiment 9 is slightly modified when the above effects are taken into account. The velocities in the suction and delivery ducts of the compressor corresponding to the test point analysed in Section 9.7.1 are respectively 60 m/s and 67 m/s (higher in the delivery duct because of the higher temperature and lower density of air). The corresponding temperature rises at the stagnation points are 1·8°C and 2·2°C. It is thus apparent that the increase in static temperature across the compressor is about 0·4°C less than the increase in total temperature, as a consequence of the higher air-velocity in the exit duct.

The experimental observations show that the temperature t_1 in the inlet duct is equal to the static temperature t_0 in the air at rest outside the machine, allowing us to conclude that in this case the mercury-in-glass thermometer in the duct is reading the true total temperature of the moving air. If the same is true of the thermometer in the discharge duct, we may conclude that the increase in static temperature across the machine is 0·4°C less than the observed difference, $t_2 - t_1 = 35·0°C$.

The increase in energy imparted to the air as a consequence of the higher velocity in the discharge duct is:

$$\Delta P_i = \tfrac{1}{2}\dot{m}(v_2^2 - v_1^2) = 0·111(67^2 - 60^2) = 99 \text{ W} \tag{10.7}$$

This correction is small when compared with the value $P_i = 7810$ W. In accurate compressor tests, measurements are made of the total pressures before and after the machine rather than the static pressures measured in Experiment 10. This renders the results independent of any difference in cross-sectional area of inlet and discharge ducts.

As is to be expected from fluid mechanics, the total pressure at the compressor inlet is very nearly equal to the external atmospheric pressure, implying that no significant losses have occurred in the course of the acceleration of the air entering the compressor through the measuring nozzle and the subsequent deceleration and pressure recovery in the inlet duct. The pressure ratio across the machine calculated on the basis of total pressure differs to a negligible degree from that calculated from the observed static pressures.

We may conclude that in the case of this machine, analysis on the basis of a control volume drawn as in Fig. 9.7.4 to exclude the inlet nozzle, and on observations of static rather than total pressures in the inlet and discharge ducts, gives results that differ to a negligible extent from those yielded by a more exact analysis.

10.3   Heat Transfer at High Gas Velocities

At velocities exceeding about 60 m/s compressibility effects become of significance in gas flow and heat transfer phenomena. In the theoretical treatment these effects are taken into account by the introduction of an additional dimensionless group, the Mach Number, representing the ratio between the gas velocity and the local velocity of sound, a.

$$(Ma) = \frac{v_x}{a}$$

The Nusselt Number becomes:

$$(Nu) = f((Re), (Pr), (Ma))$$

In the previous Section it has been shown that at high gas velocities the temperature at the stagnation point may substantially exceed that in the free stream. We shall now consider the situation in the boundary layer at high velocities. The simplest case is represented by the flat plate investigated in Experiment 1. Fig. 10.3.1 represents an analogous situation: a thin flat plate with identical flow conditions on both surfaces. This may be regarded as a simplified one-dimensional representation of the conditions encountered by a supersonic aircraft in flight. While at low velocities the only effect of friction in the boundary layer is to give rise to a tangential drag force at the plate surface, at high velocities the heat generated by these friction forces also becomes significant.

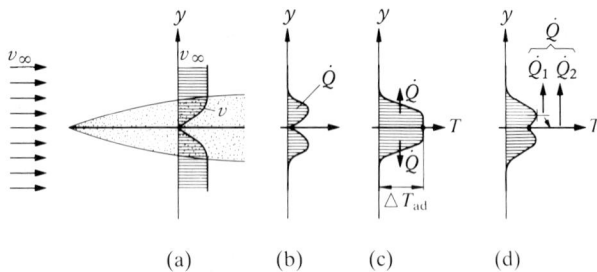

(a)   (b)   (c)   (d)

Fig. 10.3.1 High-velocity flow past a flat plate:
    (a) Velocity distribution in boundary layer (thickness exaggerated)
    (b) Temperature distribution in boundary layer, plate at same temperature as stream
    (c) Plate at same temperature as boundary layer
    (d) With some heat transfer to plate

Fig. 10.3.1(a) shows the velocity distribution in the boundary layers on each side of the plate (the thickness of the layer in the direction normal to the flow is exaggerated). Fig. 10.3.1(b) shows the temperature distribution in the boundary layer with plate temperature equal to the static temperature of the airstream. To an appropriate scale the area beneath the curve represents the total rate of heat evolution in the boundary layer. Fig. 10.3.1(c) shows the temperature distribution at equilibrium conditions, with plate temperature equal to the boundary layer temperature in immediate contact with the plate. In these circumstances there is, of course, no heat transfer from boundary layer to plate.

The value of the equilibrium temperature T_e is principally a function of the viscosity and conductivity of the gas (the viscosity influences the rate of generation of heat in the gas and the thermal conductivity its capacity for transporting that heat). To a lesser extent T_e is influenced by the Reynolds Number. For a Prandtl Number $(Pr) = 1$ the equilibrium temperature is equal to the stagnation temperature given by equations (10.5) or (10.6). For other Prandtl numbers the equilibrium temperature is related to the stagnation temperature by a recovery factor r:

$$r = \frac{\varDelta T}{\varDelta T_{Ad}}$$

For the laminar boundary layer, theory suggests a value $r = \sqrt{(Pr)}$. Fig. 10.3.2 shows experimental values of r plotted against Reynolds Number. It is apparent that the equilibrium temperature is not much below the adiabatic stagnation temperature.

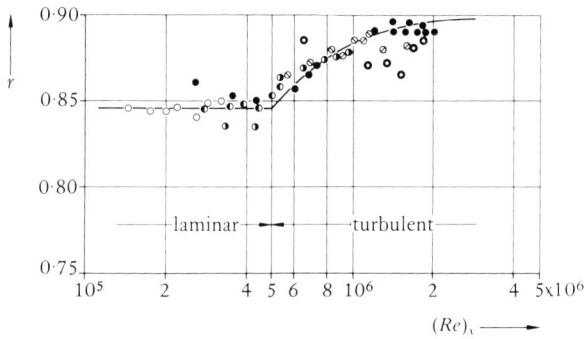

Fig. 10.3.2   Relation between recovery factor and Reynolds Number, boundary layer on flat plate

Fig. 10.3.1(d) represents the case where the plate is not adiabatic but is losing heat, for example by radiation to the surroundings. The temperature of the plate surface is then less than the maximum temperature in the boundary layer, with consequent heat transfer from the air to the plate. We thus have the surprising situation that heat is transferred from the gas stream to the plate, even though the static temperature of the gas in the region beyond the boundary layer is lower than the temperature of the plate. Part of the heat generated in the boundary layer is transmitted to regions of the boundary layer more remote from the plate surface while part is transmitted to the plate, from which it is removed by radiation or by internal conduction.

In extreme cases, for instance in the re-entry of space vehicles, stagnation temperatures can be so high that material can be sublimated from the solid surface, the heat absorbed by this process serving to limit the temperature reached by the surface to a value less than the stagnation temperature.

10.4   Further Comments on the Second Law; Entropy, a Property

So far we have assumed without proof that entropy is a property. We shall first show that this is the case for a perfect gas, and then for all substances. The First Law, equation (2.5), may be written in differential form:

$$dU = dQ - dW \tag{10.8}$$

dQ signifies here, as elsewhere in the book, not the differential of a property but the quantity of heat entering the system in time dt:

$$dQ = \dot{Q}\,dt$$

The element of work dW has an analogous significance.

Equation (10.8) understood in this way applies for any working fluid in a closed system. Let us now limit the work performed to displacement work $dW = p\,dV$. If we now wish to calculate Q_{12} from equation (10.8) corresponding to a change from state

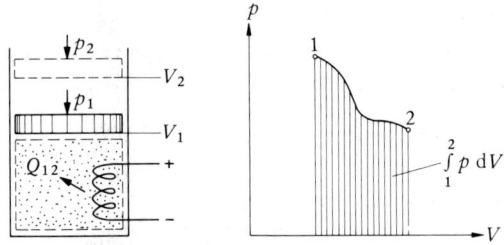

Fig. 10.4.1 Displacement work with the addition of heat

1 to state 2, Fig. 10.4.1, it is necessary for us to know not only the conditions at points 1 and 2 but the exact course of the process between these two points:

$$\int_1^2 dU = \int_1^2 dQ - \int_1^2 p \, dV; \quad \int_1^2 dQ = Q_{12} = U_2 - U_1 + \int_1^2 p \, dV \qquad (10.9)$$

Q_{12} is, unlike U, not a property. The change in the latter, $U_2 - U_1$, may be written immediately since the integral is independent of the route followed by the process: this is an essential characteristic of a thermodynamic property.

Now limit the consideration to a perfect gas for which

$$pV = mRT \qquad (10.10)$$

Here T is the Kelvin temperature defined by the gas thermometer (see p. 15). We now divide equation (10.8) by T and make use of the relations:

$$p = \frac{mRT}{V} \quad \text{and} \quad dU = mc_v \, dT$$

$$\frac{dU}{T} = \frac{dQ}{T} - \frac{p \, dV}{T}$$

$$\frac{dQ}{T} = mc_v \frac{dT}{T} + mR \frac{dV}{V} \qquad (10.11)$$

Integrating this expression between state 1 and state 2:

$$\int_1^2 \frac{dQ}{T} = mc_v \int_1^2 \frac{dT}{T} + mR \int_1^2 \frac{dV}{V} = mc_v \log_e \frac{T_2}{T_1} + mR \log_e \left(\frac{V_2}{V_1} \right) \qquad (10.12)$$

Evidently the right-hand side of this equation is integrable in the absence of knowledge of the particular path followed by the process between the two end-states. Since the right-hand side of the equation is a function of properties of the working fluid that are independent of the path of the process, the left-hand side must similarly be independent:

$$\int_1^2 \frac{dQ}{T} = s_2 - s_1$$

Thus, while Q_{12} is not a property, the integral of dQ/T is a property: entropy.

Let us bring together the assumptions made in the above:

(a) Closed system, displacement work only.

216

(b) Working fluid perfect gas, c_v constant.

(c) Change of state from point 1 to point 2 quasi-static (reversible) since only then may we assume $pV = mRT$.

These conditions, and particularly the second and third, appear at first to be very restrictive: all we have done so far is to show that in the case of a perfect gas, if such a substance existed, entropy would be a property.

The last assumption is also important. Suppose, for example, that the heat Q_{12} is imparted to the gas by passing an electric current for a brief period through a resistance immersed in the gas. The resistance wire and the gas in the boundary layer in immediate contact with it assume a high temperature, and during the process no specific temperature T can be ascribed to the gas; equation (10.11) loses its validity. The heat must be added so slowly that at any stage of the process a uniform temperature may be ascribed to the gas (a quasi-static change of state).

This conclusion is modified in the light of the following considerations. If the heat Q_{12} is added as suggested above, equilibrium conditions (p_2, U_2, T_2, s_2) will in due course prevail. We can, however, also reach state 2 through the quasi-static addition of heat, and we have already shown that the change in entropy from state 1 to state 2 is independent of the course of the intervening process. Condition c is thus only necessary to permit calculation of the entropy at state 2. The actual process between state 1 and state 2 may take any desired course.

The fact that entropy is a property permits us to construct diagrams, such as the T-s diagram, in which entropy forms one of the axes. As already discussed in Chapter 5, the relation $dQ = T \, ds$ means that on this diagram heat added to or removed from the fluid may be represented by the area beneath the curve in Fig. 10.4.2. It may, of course, be immediately deduced from this that for the Carnot cycle with a perfect gas as working fluid the efficiency is given by Fig. 10.4.3.

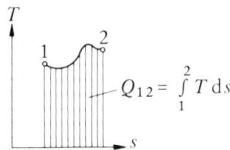

$$Q_{12} = \int_1^2 T \, ds$$

Fig. 10.4.2 Representation of heat addition on the T-s diagram

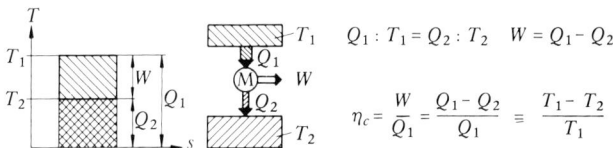

$$Q_1 : T_1 = Q_2 : T_2 \qquad W = Q_1 - Q_2$$

$$\eta_c = \frac{W}{Q_1} = \frac{Q_1 - Q_2}{Q_1} = \frac{T_1 - T_2}{T_1}$$

Fig. 10.4.3 The Carnot cycle on the T-s diagram

217

It remains to be shown that entropy is also a property of any substance and not merely of a perfect gas. Fig. 10.4.4 shows a hypothetical arrangement by which the truth of this proposition may be established. We imagine two closed adiabatic systems, one containing a real working fluid and the other a perfect gas. Each system is closed by a movable piston and the two are separated by a perfectly conducting wall. Suppose that the perfect gas is taken from state 1 to state 2 exactly as postulated at the beginning of this section (Fig. 10.4.1), except that the heat Q_{12} transmitted to the perfect gas is supplied in a quasi-static manner from the real fluid. If we compress the real fluid by an infinitesimal amount, its temperature will rise and heat be transferred through the conducting wall to the perfect gas. We can regulate the movement of the piston enclosing the real fluid so that heat transfer takes place at the appropriate rate to the perfect gas. We have already shown that for the latter the change of entropy is independent of the course of the process and is given by equation (10.12).

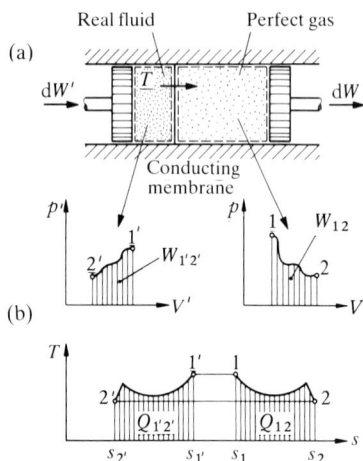

Fig. 10.4.4
(a) Adiabatic closed systems for a real fluid and a perfect gas with heat transfer between them
(b) p-V and T-s diagrams

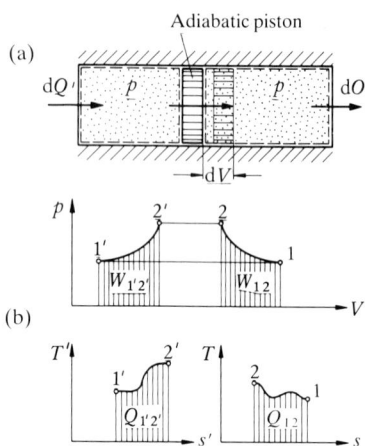

Fig. 10.4.5
(a) Adiabatic closed systems for a real fluid and a perfect gas with frictionless adiabatic piston between them
(b) p-V and T-s diagrams

The T-s diagram for the perfect gas and the real fluid are mirror images of each other: each successive increment of heat dQ is transmitted to the perfect gas at the same temperature as that at which it leaves the real fluid, and $ds = dQ/T$ is identical for both fluids. Since the change of entropy from state 1 to state 2 for the perfect gas is independent of the path of the process, we have now shown that the same applies for the real fluid, of which entropy is also therefore a property. The Kelvin temperature scale, which was originally defined on the basis of the perfect gas, attains universal significance, and the Carnot cycle efficiency is identical whatever the working fluid.

The arrangement shown in Fig. 10.4.4 suggests an analogy between heat and work. We postulate two adiabatic systems, one containing a real fluid and the other a perfect gas, separated by a non-conducting piston. While in the case of the arrangement of Fig. 10.4.4 the T-s diagrams for the two fluids are mirror images and the p-V diagrams differ, in the case of Fig. 10.4.5 the p-V diagrams are necessarily mirror images but the T-s diagrams are not.

Instead of a single real fluid we may suppose the left-hand working space in Fig. 10.4.4 to be occupied by a system of any kind we care to propose. If this system exchanges heat with the ideal gas in a quasi-static manner, its change in entropy is defined exactly as was the case with a single fluid. If within this system an electric motor were to perform mechanical work, the entropy of the system would remain constant in the absence of losses but would increase if loss were present. In order to calculate the increase of entropy we must devise a series of reversible processes that will arrive at the same end condition:

(a) Transformation of an amount of electrical power equivalent to the mechanical work performed with 100 percent efficiency.

(b) Transfer of a quantity of heat from the ideal gas to the system equivalent to the electrical power-loss in the motor. Increase in entropy is then

$$s_2 - s_1 = \frac{W_1}{T} \quad \left(\text{or} = \int_1^2 \frac{\mathrm{d}W_1}{T} \quad \text{if } T \text{ is not constant} \right)$$

where $W_1 = $ motor losses.

As entropy is a property of a working fluid we can naturally apply the above arguments to steady-flow processes taking place in open systems, for example to a steam cycle.

It will be apparent that in an adiabatic closed system in which only reversible processes are taking place the entropy will remain constant since

(a) No heat is exchanged with the surroundings.

(b) By definition, transformations of chemical, electrical, mechanical and potential energy take place without loss (i.e. without the evolution of heat).

(c) Also by definition, any heat engine cycles operating within the system are reversible and, like the Carnot cycle, result in no increase in entropy.

It should be noted that in this discussion of the concept of entropy as a property and its relation with the Second Law we have started by postulating the existence of an ideal gas having constant c_v. Without this hypothesis the argument becomes significantly more difficult and abstract. It is, however, in any case necessary to postulate the existence of an ideal gas in the derivation of the thermodynamic or Kelvin temperature scale. Monatomic gases at moderate pressure and temperatures well above the critical behave almost exactly as ideal gases.

10.5 Kinetic Theory of Heat

In Chapter 2 certain relationships between molecular velocities, pressure and temperature in a gas, were given without proof.

From the point of view of the kinetic theory an ideal gas is regarded as an assembly

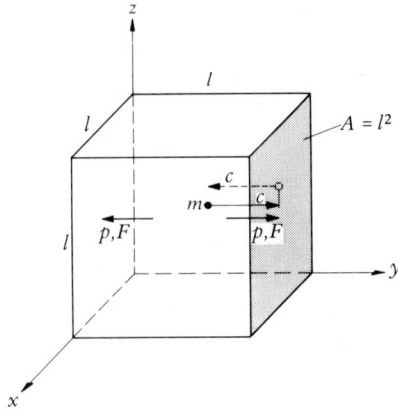

Fig. 10.5.1 Development of gas pressure in accordance with the kinetic theory

of point masses between which, except at impact, no forces are exerted. The impact of these bodies on the walls of the container gives rise to the gas pressure. Fig. 10.5.1 represents a cubical container of side length l containing N molecules of individual mass m. The true distribution of velocities and directions of motion of the molecules is regarded as equivalent at any instant to the motion of $N/3$ molecules at an average velocity c in the x, y and z directions. Each molecule is regarded as travelling between opposite walls at velocity c and will thus make $c/2l$ impacts per second on each wall. The total number of impacts per second on the wall will then be:

$$\dot{n} = \frac{Nc}{3 \cdot 2 \cdot l}$$

At each impact the momentum of the molecule changes from

$$J_1 = +mc \quad \text{to} \quad J_2 = -mc, \quad \text{or} \quad J_1 - J_2 = 2mc.$$

For $\dot{n}$ impacts the corresponding force is given by

$$F = \dot{n}(J_1 - J_2) = \frac{Nc}{6l} \cdot 2mc = \frac{Nmc^2}{3l}$$

The pressure exerted is therefore

$$p = \frac{F}{A} = \frac{Nmc^2}{3l^2} = \frac{1}{3}\frac{M}{V}c^2 = \frac{1}{3}pc_2$$

or

$$c^2 = \frac{3p}{p} = 3pv = 3RT$$

Here c^2 is a mean square velocity, the root of which is only approximately equal to the mean velocity of the molecules. Fig. 10.5.2 shows the true distribution of molecular velocities for nitrogen at two different temperatures. This distribution follows the Maxwell–Boltzmann distribution function [3]. An exact calculation, taking into account this velocity distribution and the fact that in reality the molecules will be

220

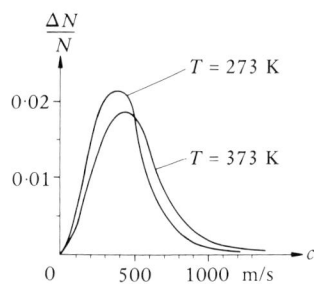

Fig. 10.5.2 Distribution of molecular velocities for nitrogen

travelling in all possible directions leads to the following expression for the mean molecular velocity,

$$c_m = \sqrt{\frac{8RT}{\pi}} \qquad (10.13)$$

Appendix

Laboratory Practice and Experimental Method

The purpose of this Appendix is to give a brief review of the various aspects of laboratory practice that must be mastered by a competent experimental engineer.

The purpose of an experiment is to provide the answer to specific questions. Typical questions concerned with the subject-matter of this volume would be:

(a) The dimensional analysis of forced convection leads us to believe that this will depend on three dimensionless groups, the Nusselt, Prandtl and Reynolds Numbers. Test the validity of this theory by making measurements of the heat transfer from a heated cylinder to a transverse flow of air. Do you think that the apparatus available permits the exploration of a sufficiently wide range of variables, and would you expect other features, not predicted by the theory, to emerge under extreme conditions?

(b) We have in the laboratory a small reversible air-conditioner/heat pump. This is intended for multiple installation in large buildings where there may in some areas be excess heat and in others a lack of heating. The idea is that in the hot areas units should be installed to act as coolers, transferring the resultant heat in the form of hot water to other units acting as heat pumps installed in the cold parts of the building.

Examine the performance of the air-conditioner/heat pump and assess the economics of this process.

(c) Produce a complete characteristic showing the relationship between pressure ratio, mass flow, efficiency and surge limit over a range of operating speeds for a centrifugal compressor.

Question (a) is of fundamental nature and could give rise to an extensive programme of research; nevertheless it is clearly defined and specific. A laboratory experiment intended to answer at any rate part of this question is more likely to produce meaningful answers than an experiment that is a response to the introduction "study the phenomenon of forced convection".

Question (b) calls for an experimental programme of a more practical nature than that called for by question (a), but the experimenter will, nevertheless, need to consider a number of questions before he can start work in the laboratory.

For example:

(i) To what range of operating conditions must we subject the equipment in order to obtain sufficiently comprehensive data? How are we to obtain these variable test conditions?

(ii) Are the methods of measurement available, and in particular the means for measuring the rate of flow of air through the apparatus, sufficiently accurate to give reliable results?

(iii) How do we decide whether the particular air-conditioner/heat pump that we are using is a good example of the kind of apparatus or not?

Against what standard do we assess its performance and what measurements do we need to take in order to distinguish between different aspects of this performance and to decide where shortcomings, if any, may lie?

Question (c) is a typical example of an industrial test programme intended to establish the performance of a product. It does not call for a deep study of the underlying theory, which is not the concern of the experimenter on this particular occasion, though questions of accuracy will be of prime importance. In planning this experiment careful consideration must be given to the test conditions, for example to the positioning and calibration of the flow-measuring device and to the accuracy of the associated instruments for measuring pressures and temperatures.

Bodies such as the International Standards Association publish detailed instructions for carrying out this kind of experiment, and the engineer should be familiar with the standards applicable to his own area of activity and should gain experience in the detailed application of these standards.

The design of experiments is a difficult art, best learned in the early stages by carrying out, with a critical mind, experiments designed by others.

The end-product of the work in the laboratory is a report, the purpose of which is to give answers to the questions that were asked and to explain the significance of these answers. In some cases the report will be intended for the experimenter himself, but in most it will be intended for someone else.

We can now define the experimenter's task. He must:

(a) Know and understand the questions that his experiment is intended to answer and the needs of the man who has asked them.

(b) Have a sufficient understanding of the relevant theory.

(c) If it does not already exist, design and assemble the necessary apparatus.

(d) Carry out the experiment, ensuring that the accuracy of his results is sufficient to meet the requirements of the questioner.

(e) Reduce the experimental readings and present them in a suitable form and with the accuracy required.

(f) Summarize the experimental findings, relating them to the question asked at the beginning.

(g) Discuss the findings and satisfy the reader that the answers given are correct.

Formal reports may follow the same logical sequence:

Purpose of Experiment
Theoretical Background
Description of Apparatus and Experimental Method
Results and Calculations
Conclusions
Discussion

It is suggested that, for the majority of laboratory reports, background theory, description of apparatus and experimental method may be covered by duplicated laboratory sheets, while the results may also be recorded and reduced on proforma sheets; there is little merit in the mechanical copying out of information. The student is then required to write only the conclusions and the discussion.

In his subsequent professional life the student will, if he is involved in laboratory work at all, be required to submit reports to his superior or perhaps to professional clients. These professional reports must fulfil rather different requirements from those appropriate to the ordinary teaching laboratory report, and it is suggested that a certain number of experiments should be written up as if they were intended for this purpose. Imagine a Chief Engineer who

(a) is a busy man who requires a short clear report to answer certain specific questions;

(b) does not need a detailed account of the equipment, which he already knows, but does need a clear and accurate account of the instrumentation used and the experimental methods adopted;

(c) is concerned with the accuracy of the results and needs to be satisfied that this is adequate;

(d) needs to be convinced by an intelligent discussion of the results that the problem has been understood and that the correct answers have been given.

Reports of this kind are particularly appropriate to "project work" during the later stages of the course.

Laboratory work, at any rate in the engineering field, is usually a group activity and most of the experiments dealt with in this volume may appropriately be carried out by a group of four students working together. One student should direct the test and record the results, while the remainder of the group should each be responsible for taking certain of the necessary readings. The group should be rotated so that each member gains experience in each task. It is desirable that the results should be roughly plotted in the course of the laboratory period so that faulty readings may be rejected and any unduly large gaps between successive test-points filled in.

Laboratory work is concerned largely with the taking of measurements, and a major purpose of any laboratory course is the inculcation of skill in the use of measuring instruments. There is something to be said for preliminary courses in "instrumentation", though the authors are inclined to the view that such courses fail to capture the interest of the students and that this knowledge is best acquired incidentally in the course of general experimental work. Skill in this field comprises a number of elements:

(a) Experience in the correct use of instruments.

(b) Knowledge of methods of calibration and an awareness of the different errors to which instruments are subject.

(c) A critical understanding of the relative merits and limitations of different methods of measurement and their applicability to different experimental situations.

(d) An understanding of the differences between true and observed values of experimental quantities.

Among the more important basic skills may be mentioned the following:

(a) Correct interpolation. Fig. A.1 shows a pointer reading the value 9·3; this will commonly be read as 9·2 or 9·4.

Fig. A.1 Interpolation of gauge readings

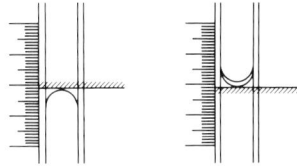

Fig. A.2 Reading of a meniscus

(b) Reading of manometers. Fig. A.2 shows a mercury meniscus and a meniscus for a fluid that wets the manometer tube. A straight-edge should be used to read the true level.

(c) Averaging of readings. This is particularly difficult in the case of dynamometer torque measurements when studying the power output of machines. Small random variations in the power output of diesel engines occur continuously and it is most desirable that, when taking a fuel consumption, the dynamometer torque reading should be observed continuously throughout the measurement and an average value taken rather than a single spot reading.

(d) Accurate checking of zero values.

(e) The recordings of readings to the correct number of significant figures. If a manometer level is oscillating with an amplitude ±5 mm it is meaningless to quote the average level to an accuracy greater than 1 mm. One of the commonest student errors is to omit the final zero when this is a significant figure. In the following column of results, for example, the third figure should be recorded as 9·0:

$$3·6$$
$$7·2$$
$$9$$
$$11·1$$

(f) The allowance of sufficient time for readings to stabilize, particular important where temperatures are concerned.

(g) The accurate use of timers. It is an instructive experiment to ask a student to make a number of successive timings of a regularly repeated process, for example, to time a period of ten seconds on a continuously running stopwatch, using a second watch for the purpose. The results are recorded by a second student without being seen by the first. A number of such successive readings will show a degree of scatter and this scatter varies considerably from one person to another.

(h) Correct use of such standard laboratory apparatus as chemical balances.

(i) The acquisition of a general skill in the handling of instruments and controls. A recognition that the calibration knob on an electrical instrument must be handled differently from the handwheel of a 10 cm hydraulic valve.

225

(j) Self-confidence and steady nerves in the handling of powerful and sometimes noisy machinery. The real engineer must be master of his creations; it is no good running away if the throttle of a petrol engine sticks in the "open" position and the engine starts to accelerate uncontrollably.

There is a strong tendency for the inexperienced laboratory worker to believe the readings of his instruments; the experienced experimenter knows that every reading is suspect, that he must calibrate his instruments, and that he must be continually on the alert to notice faults in the installation or performance of his instruments.

The introduction of digital instrumentation has, paradoxically, greatly increased the probability of uncritical acceptance of unreliable observations. This is because a digital read-out, giving an apparently quite specific numerical value to the quantity to be measured, conveys an impression of accuracy that may be in no way justified.

An analogue indicator associated with a thermocouple can in most cases not be read to closer than 1 deg. C, and the act of reading the indicator is likely to bring to mind the many sources of inaccuracy in the whole temperature-measuring system.

Replace the analogue instrument by a digital indicator reading to 0·1 deg. C and it is easy to forget that all the sources of error are still present: the fact that the read-out can now discriminate to ± 0.05 deg. C does *not* mean the overall accuracy of the reading is within comparable limits.

Engineering students should be fully aware of this dangerous feature of digital instrumentation.

A consideration of the relative merits of different types of instrumentation for measuring the same quantity is a vast subject and is beyond the scope of this Appendix. One or two principles will be mentioned:

(a) Cumulative measurements are generally preferable to rate measurements. Examples are the measurement of rotational speed by counting revolutions over a period of time, and the measurement of fuel consumption by recording the time taken to consume a given volume or mass, rather than by using a previously calibrated rate meter.

(b) "Absolute" methods are to be preferred to "relative" methods. Pressure measurement by a column of mercury is preferable to measurement by a Bourdon gauge. A spring balance is a better way of measuring a constant force than a transducer; a dead-weight system is better than either.

(c) In instrumentation, as in engineering generally, simplicity is a virtue. Each elaboration is also a potential source of error. This applies particularly to electronic apparatus.

Many different factors help to determine the difference between the true and the apparent value of any given experimental observation. As an illustration consider the arrangement shown in Fig. A.3 which represents a typical experimental situation: the determination of the temperature of the exhaust gas leaving a diesel engine. The instrument chosen is a vapour pressure thermometer, such as is suitable for temperatures of up to 600°C. It comprises a steel bulb, immersed in a gas of which the temperature is to be measured, and connected by a long tube to a Bourdon gauge which senses the vapour pressure but is calibrated in temperature.

Let us consider the various errors to which this system is subject.

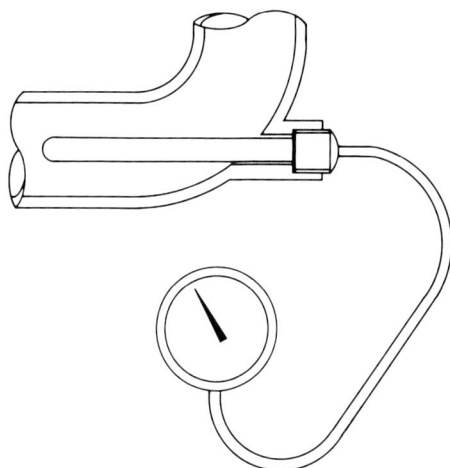

Fig. A.3   Measurement of temperature in a flowing gas

Sensing errors are associated with the interface between the system on which the measurements are to be made and the instruments responsible for those measurements. In the the present case there are a number of sources of sensing error. In the first place, the bulb of the temperature indicator can "see" the walls of the exhaust pipe and these are inevitably at a lower temperature than that of the gas flowing in the pipe. It follows that the temperature of the bulb must be less than the temperature of the gas. This error can be reduced but not eliminated by shielding the bulb or by employing a "suction pyrometer". A further source of error arises from heat conduction from the bulb to the support, as a result of which there is a continuous flow of heat from the exhaust gas to the bulb and no equality of temperature between them is possible.

A more intractable sensing error arises from the circumstances that the flow of gas in the exhaust pipe is constant neither in pressure, velocity nor temperature. Pulses of gas, originating at the opening of the exhaust valves in individual cylinders, alternate with periods of slower flow, while the exhaust will also be to some extent diluted by scavenge air carried over from the inlet. The thermometer bulb is thus required to average the temperature of a flow that is highly variable both in velocity and in temperature, and it is unlikely in the extreme that the actual reading will represent a true average.

A more subtle error arises from the nature of exhaust gas. Combustion will have taken place, resulting in the creation of the exhaust gas from a mixture of air and fuel, perhaps only a few hundredths of a second before the attempt is made to measure its temperature. This combustion may be still incomplete, the effects of dissociation arising during the combustion process may not have worked themselves out, and it is even possible that the distribution of energy between the different modes of vibration of the molecules of exhaust gas will not have reached its equilibrium value; as a consequence it may not be possible even in principle to define the exhaust temperature exactly.

Bourdon-type pressure gauges are particularly prone to a variety of *instrument errors.* Two of the commonest are *zero error* and *calibration error*. The zero error is present if the pointer does not return precisely to the zero graduation when the gauge is subjected to zero or atmospheric pressure.

Calibration errors are of two forms: a regular disproportion between the instrument indication and the true value of the measured quantity, and errors that vary in a non-linear manner with the measured quantity. This kind of fault may be eliminated or allowed for by calibrating the instrument, in the case of a pressure gauge by means of a dead weight tester. These are examples of *systematic errors.*

In addition, the pressure gauge may suffer from *random errors* arising from friction and backlach in the mechanism. These errors affect the *repeatability* of the readings.

Further to these various sources of inaccuracy we have *observer errors.* The magnitude of these depends partly on the skill of the observer; reference has been made above to errors in estimating the position of a pointer, and further errors may arise due to parallax effects (the pointer not being viewed normal to the dial surface) and to carelessness or error in recording the readings.

The *sensitivity* of an instrument may be defined as the smallest change in applied signal that may be detected; in the case of a pressure gauge it is affected particularly by friction and backlash in the mechanism. The *precision* of an instrument is defined in terms of the smallest difference in reading that may be observed. Typically, it is possible to estimate readings to within 1/10 of the space between graduations, provided the reading is steady, but if it is necessary to average a fluctuating reading the precision may be much reduced.

Finally one must consider the effect of *installation errors.* In the present case these may arise if the bulb is not inserted with the correct depth of immersion in the exhaust gas or, as is quite often the case, if it is installed in a pocket and is not subjected to the full flow of the exhaust gas.

A consideration of this catalogue of possible errors will make it clear that it is unlikely that the reading of the indicator will reflect with any degree of exactness the temperature of the gas in the pipe. It is possible to analyse the various sources of error likely to affect any given experimental measurement in this way, and while some, for example measurements of length and weight, require a less complex analysis, others, notably readings of inherently unsteady properties such as flow velocity, require to be treated with scepticism. A hallmark of the experienced experimenter is that, as a matter of habit and training, he questions the accuracy and credibility of every experimental observation.

It is usual to include some consideration of the mathematical theory of errors in an engineering course, and the reader is referred to standard texts for a treatment of this subject. It will only be remarked here that it is extremely difficult to place a definite tolerance on the accuracy of any experimental observation or even to say precisely what that tolerance means. It is generally assumed that if a large number of observations of a nominally constant quantity are made, those observations will form a Gaussian distribution. We may calculate the standard deviation σ for such a distribution and predict that 95% of the readings will lie within the limits (*the probability zone*) $\pm 2\sigma$ and 99·7% within the limits $\pm 3\sigma$. Strictly speaking, when defining the *limits of error* of our observations we should also define the probability zone to which these limits refer. In any case these limits only cover the random errors and do not deal with the systematic errors which, in the case of the example given above, outnumber the sources of random error.

Every effort must be made to eliminate systematic errors by calibration of the instrument and by the avoidance of faults in its installation. Only when we are satisfied that errors of this kind have been eliminated can we deal with the random errors by taking a number of readings and analysing these statistically, allowing us to specify a probability zone within which the true value is likely to lie.

A common error in presenting experimental results is the addition of meaningless significant figures when reporting derived quantities. Consider, for example, the equation for air flow quantity in terms of pressure drop across a measuring device. This might take the form for given conditions of temperature and pressure:

$$\dot{m} = 0{\cdot}04081\sqrt{h} \quad \text{kg/s}$$

Now suppose that the head h is read as:

$$100 \pm 1 \text{ mm } H_2O$$

We could calculate that the corresponding mass flow $\dot{m} = 0{\cdot}4081$ kg/s. However, the tolerance on our reading is such that the true velocity head could have any value between 99 and 101 mm H_2O; the corresponding quantities are $0{\cdot}4061$ and $0{\cdot}4101$ kg/s. It is thus apparent that we must apply a tolerance of $\pm 0{\cdot}002$ kg/s to our calculated quantity and that the last significant figure in our value $\dot{m} = 0{\cdot}4081$ is quite meaningless.

A rather similar pitfall occurs when we multiply together two or more observed quantities, one of which is known to a less degree of accuracy or to a smaller number of significant figures than the others. In these cases it is useful as a check to take a set of values lying at each end of the limits of probable error for each of the component quantities and to calculate the corresponding spread in the resulting product. This will lead to a realistic assessment of the number of significant figures to which the product may justifiably be reported.

The quality of a piece of experimental work depends to a considerable degree on the accuracy and reliability of the measurements made; a large part of the art of the experimenter lies in his ability to look at his measurements critically and to ensure that they accurately represent the quantities he is attempting to observe.

References

[1] ROGERS, G. F. C. and MAYHEW, Y. R. (1980, 3rd edn) *Engineering Thermodynamics Work and Heat Transfer*. Longman, London. A comprehensive text at undergraduate level.

[2] WALLACE, F. J. and LINNING, W. A. (1970, 2nd edn), *Basic Engineering Thermodynamics*. Pitman, London. A similar text to [1]. A particularly clear treatment of availability.

[3] HATSOPOULOS, G. N. and KEENAN, J. H. (1965), *Principles of General Thermodynamics*. Krieger, Melbourne, Florida. A classic text with emphasis on the historical and philosophical aspects of the subject.

[4] MCADAMS, W. H. (1973, 3rd edn), *Heat Transmission*. McGraw-Hill, London. A comprehensive text.

[5] ROHSENOW, W. M. and HARTNETT, J. P. (1985, 2nd edn), *Handbook of Heat Transfer*. McGraw-Hill, N.Y. A clear account of recent developments.

[6] HAHNE, E. and GRIGULL, U. (1977), *Heat Transfer in Boiling*. Hemisphere, London.

[7] LOVE, T. J. (1968), *Radiative Heat Transfer*. Merrill, Columbus, Ohio.

[8] SHEPHERD, D. G. (1960), *Introduction to the Gas Turbine*. Constable, London.

[9] RICARDO, H. R. (1968), *The High Speed Internal Combustion Engine*. Blackie, London. A classic, notable for its clarity and insight.

[10] GREENE, A. B. and LUCAS, G. C. (1969), *The Testing of Internal Combustion Engines*. E.U.P., London.

[11] PLINT, M. A. and BÖSWIRTH, L. (1978), *Fluid Mechanics: A Laboratory Course*. Griffin, London. A companion text to the present volume.

[12] BS 1041: 1943. Code for Temperature Measurement. (Withdrawn and replaced by specialist Parts 2 to 7.)

[13] BS 1041: Section 2.1: 1985. Liquid-in-Glass Expansion Thermometers.

[14] BS 1041: Part 4: 1966. Thermocouples.

[15] BENEDICT, R. P. (1984, 3rd edn), *Fundamentals of Temperature, Pressure and Flow Measurements*. Wiley, N.Y.

[16] HEYWOOD, H. (1954). Solar Energy for Water and Space Heating. *J. Inst. Fuel*, **27**, 162.

[17] *Department of Energy Paper No. 16* (1976), Solar Energy, its Potential Contribution within the United Kingdom. H.M.S.O.

[18] KEENAN, J. H. *et al.* (1978), *Steam Tables (S.I. Units)*. Wiley, N.Y.

[19] KEENAN, J. H. *et al.* (1983, 2nd edn), *Gas Tables*. Wiley, N.Y.

[20] *Refrigerants: Pressure Enthalpy Diagrams*. Mond Division, ICI Ltd, London.

[21] *Psychrometric Charts*. CIBSE, London.

[22] *Psychrometric Tables: Guide Book C* (1975), Fig. 1/2. Properties of Air, Water and Steam. CIBSE, London.

[23] BS 1042: Part 1: 1981/84. Orifice Plates, Nozzles and Venturi Tubes.

[24] KASTNER, L. (1947). The Airbox Method of Measuring Air Consumption. *Proc. I. Mech. E.*, **157**, 367.

Index